System and Systems Thinking

Fundamental Theory and Practice

(Core Book with Extra Teaching Material)

International Easy English Edition

A. Gharakhani Bahar

Bibliographic Information

ISBN:
978-91-981875-0-2

Title:
System and Systems Thinking - Fundamental Theory and Practice
(Core Book with Extra Teaching Material)

Author:
A. Gharakhani Bahar (1951-   )
a_gh_bahar@hotmail.com

First Publication:
May, 2014

Keywords:
System, Systems Thinking, World, Objects, Events, Order, Rule, Structure, Behavior,
Discipline, Matter, Energy, Information, Structural Balance, Behavioral Equilibrium, Disciplinal
Certainty, Stability, Entropy

Author's Note

This paperback format includes the main book "System and Systems Thinking - Fundamental Theory and Practice" and its extra section. Extra section includes more teaching material for the instructors and students who wish to use the book in a university interdisciplinary or organizational vocational course. Main book and this extra section are also available in Kindle format on Amazon.

In the book, the most basic and fundamental concepts and methods of system and systems thinking that are common in all fields of the human knowledge, are presented in the form of a fundamental framework in theory and practice. This fundamental framework can be useful in dealing with the things in the world around from a new sight. A full list of references of the book is presented at the end of the book.

Book has gathered and presented related fundamental concepts and methods of system and systems thinking. As you will discover, these concepts and methods have completed each other as the pieces of a puzzle as a whole in a concrete and interesting content. Readers may reach the governing "spirit" of the whole content of the book in the second or third time of reading it carefully. Book also includes a rich glossary of the related concepts.

Author would be pleased to receive the comments of the readers about the book. If there is any idea about the book or its content, readers can leave comments on the related web pages in the internet. Readers also can use the following email address to send their comments directly to the author. All comments are welcome!

A. Gharakhani Bahar
a_gh_bahar@hotmail.com

Table of Contents

-1

Introduction to the Book

Introduction
The goal of this introductory chapter organized in 2 sections, is to review the backgrounds of the subject related with "system" and "systems thinking". Section one is a general review of the subject and section two presents some historical backgrounds. Indeed, what caused to write this book is what introduced at the bottom of the first and in the whole of the second section in this chapter.

Subject Background General Review
During the 20th century, many efforts were made to find a general scientific method, useable in different fields of scientific studies. The outcome of these efforts was the "General Systems Theory = GST" introduced in 1945. Although the first scientists worked in this way were from natural sciences field, but the outcome of their work went far from their fields and affected the nature of the human thinking method in general.

Among the many scientists worked in this way, were Alexander Bogdanov (1873-1928), medic from Russia, Ludwig Von Bertalanffy (1901-1972), biologist from Austria, Norbert Wiener (1894-1964), mathematician from USA, Ross Ashby (1903-1972), psychiatrist from UK, Kenneth Boulding (1910-1993), economist from UK and Russell Ackoff (1919-2009), organization theorist from USA.

The concepts of system or systems thinking first were introduced by Bogdanov at the first decades of 20th century. Bogdanov in his book "The Universal Science of Organization"[1], published in 1912, described his ideas about creating a general science for organization and called it "tectology". But the real introducing of the subject was by Bertalanffy. Bertalanffy published his ideas about systems in his book "General Systems Theory"[2] in 1945.

Bertalanffy defines "system" as a whole consisted of related parts and different from the sum of its parts. According to Bertalanffy, in different fields of science, always there have been and also there are tendencies for coming together and some kind of unification. As Bertalanffy says, the general systems theory can help the unification of various fields of science.

---

[1] The Universal Science of Organization, Bogdanov, A. A., 1922, Moscow (in Russian), Translated and reprinted by Dudley, P., 1996, Bogdanov's Tektology, Centre for Systems Studies, University of Hull.
[2] General Systems Theory: Foundations, Development, Applications (An English Language Edition), Von Bertalanffy, 1968, New York.

Bertalanffy as a biologist found that the living creatures are not only depended on their organs as their constituent parts, but also are as wholes with specific organization, different from the sum of the organs. He considered them as systems. Bertalanffy also confirmed that like the living creatures, also the world has its own organization and we are faced with related instead of unrelated objects.

According to the Bertalanffy, living creatures in order to continue their living need to interact with their environment. Indeed, stop of interactions with the environment, finally leads to the death of the living creature. He called them as "open systems". Bertalanffy also considered the systems with no interactions with their environment, as "closed systems".

Since in the open systems the environment has important role in the function or behavior of system, open systems have their own protecting and controlling mechanism and the ability of adapting themselves with the environment. But, in the case of closed systems since there is not any interaction with the environment, so there is not such a mechanism and the ability of adapting.

Following the works in the field, in 1948 Wiener published his book "Cybernetics: Control and Communication in the Animal and Machine"[3] and introduced the new science of "Cybernetics". Based on the title and content of this book, the subject of Cybernetics had been defined as control and communication in the animal and machine with aid of "information". As we will see in the next chapters, Cybernetics added the new main resource "information" to the two older main resources "matter" and "energy" in systems.

In other words, the subject of Cybernetics as a complement to general systems theory is information based control and communication with aid of "information feedback loops". This mechanism is widely used in many systems from the mechanical "hard systems" like technological devices to the social "soft systems" like animal or human kind societies.

The word "Cybernetics" has its root from the Greek word "kybernetes" as "steersman" or "steering wheel of ship". Based on this, Cybernetics as the "science of the guidance and steering", talks about the "steering" in systems. With the advent of information networks and Internet in the recent years, the word also has been used to point to information space as the "cyber space".

As a result of more discussions about the systems, in 1956 Boulding published his article "General Systems Theory: The Skeleton of Science"[4]. In his article, Boulding presented his ideas about the systems and described the methods according to that general systems theory could find its own structure.

---

[3] Cybernetics: Control and Communication in the Animal and Machine, Wiener, N., 1948, Cambridge, Mass.
[4] General Systems Theory: The Skeleton of Science, Boulding, K., *MANAGEMENT SCIENCE*, Apr. 1956, pp. 197-208.

According to the Boulding, for doing this, first method is to find the phenomena common in the experimental world that can cover more fields. Then we must try to make general theoretical models for these phenomena. The second method is to classify and arrange experimental fields in a hierarchy of their organizational complexity.

Boulding points to some common phenomena like birth and death or growth in human or animal societies that have their own certain definitions and then classifies different kinds of systems. Boulding classifies systems in three levels and each level in three classes and at sum all the systems in nine classes.

In Boulding's classification of systems, the first level that is called "Physic-o-mechanical systems" includes the 3 classes of "static structures", "simple dynamic systems" and "control mechanisms\ cybernetic systems". The second level that is called "biological systems" includes the 3 classes of "open systems\ self-maintaining structures", "genetic-societal systems" and "animals". The third level that is called "social systems" includes the 3 classes of "human", "social organizations" and "transcendental systems".

Based on this and the Bertalanffy's definition of open and closed systems, only the first level systems are as "closed" systems and the systems of other two levels are as "open" systems. About the name of the last or 9th class (transcendental systems), Boulding notes that we can not say that all the systems are these and there is not any other system. But it is possible that in the future new systems may be identified as transcendental when compared with the lower class systems. We put them at the bottom of our classification.

The next work in the field of systems again was in Cybernetics. In 1957, Ashby published his book "An Introduction to Cybernetics"[5] and explained some concepts about systems. The title of the second chapter of this book was "change" and the fifth chapter "stability". Indeed, while the "change" is in the nature of most systems, achieving a situation as "stability" that all systems tend to achieve is also desired. In other words, while the "change" is necessary, coming to "stability" and preserving it is also necessary too.

After near to 25 years of study in the field of systems, there was not yet a framework or a concrete set of words or glossary to describe the basic concepts of systems. In 1971 Ackoff published his article "Towards a System of Systems Concepts"[6] and tried to describe the concepts related to the systems in the form of system concept.

In the introductory part of this article, Ackoff notes that, since the general systems theory is a general theory, many researchers use it in many different fields. Researchers, due to the differences in their work fields are far from each other and in contrast to the key concept of system that is having relations, have no relations with each other.

---

[5] An Introduction to Cybernetics, Ashby, R., 1957, London.
[6] Towards a System of Systems Concepts, Ackoff, R. *MANAGEMENT SCIENCE,* Jul. 1971, pp. 661-671.

According to the Ackoff, in this condition, there is not a concrete or solid set of systems concepts even about the concept of system itself. In consequence, in practice various words are used for the same concept or various concepts are expressed with the same word.

Later, Ackoff notes the decade of 1940 as the end of "machine age" and the beginning of "system age". "Machine age" with emphasizing on the "parts" began with the industrial revolution. Then, system age replaced the "parts" with the "whole". In consequence, in studying the objects, instead of the "parts", the "whole" was at the center of attentions. In this decade, many philosophers, mathematicians, biologists and other scientists had worked on this subject, but apparently the creation of what Ackoff points in his article (mentioned above), had not yet occurred.

Today, searching the systems and systems thinking keywords in the internet shows that the books or articles mentioned above, even after decades of their first publication, are referred for the basic and fundamental concepts in this field. This shows that what Ackoff points in his article, still remains and needs more attentions.

At first, only the hard physical systems were at the center of attentions of the researchers. But later, also the soft logical systems attracted the attentions. Hard systems are sensible. But soft systems are understandable. In soft systems, unlike the hard systems, elements, relations among and the boundary of the system are not clear and can not be identified easily as in hard systems.

Indeed, the advent of the soft system concept was the result of efforts made to approach complex situations. Because of the ambiguity of conditions and unidentifiable interrelations among the elements in soft systems, usually there is not a clear and formal definition about these systems and any interpretation in one side may affect the case in the other side.

Among the other researchers later worked on systems and developed the concepts in this field, the following researchers can be noted: Anatol Rapoport (1911-2007) from Russia, West Churchman (1913-2004) from USA, Jay Forrester (1918) from USA, Peter Checkland (1930) from UK, Barry Richmond (1947-2002) from USA and many others.

By the works in this field, "system" as a word and concept, gradually and widely expanded not only in the scientific communities, but also in the every day life conversations of many people in many languages. Now, the word is so common that many people use it in every day life several times to describe the thing that he or she has in his or her mind.

Now, in every day life we are faced with the statements like "system does not work properly", "system failed to function", "the fault is from system side" and many others. Finally, we put ourselves out of the problems and the cause of all the faults or failures from an unknown thing named "system" that in practice we do not know any more about it!

But, really what is "system" that we talk about it so many in our every day life? Indeed, even today, not only about related concepts, but also about the system itself, there are not exact definitions. This book is a small step in this way in order to introduce some basic definitions in systems field.

Some Historical Backgrounds

Man, from the time of coming into existence on the earth, besides searching for food and protecting himself from the harms of the world around, always has tried to identify the objects and its governing orders and their related events and its governing rules. In order to respond to this request, man gradually has established various sciences.

For man, at the first glance, objects have appeared in their fixed external shapes in the space. We can say that the advent of geometry as one of the oldest sciences was due to the demand of identifying the objects and its governing orders. Also events at the first glance have appeared in the form of changes by time in their related objects in space. Again it can be said that the advent of chronology as one the oldest struggles for identifying the events, was due to the demand of identifying the events and its governing rules.

According to the above, geometry and chronology can be considered as two of the first products of human mind in identifying the objects and their related events in the world around. At first, objects and events considered separately. In other words, geometry dealt with the objects and chronology dealt with the events out of the objects itself. Relating objects and events to each other, for example in the form of cause and effect process, has occurred later.

It may be very amazing for us, while the modern world has so evolved and has changed enormously from what it was in the ancient, how the Euclidean Geometry is also valid and is used today. Also, it may be amazing when we think that in the time that there has not been any remarkable advancement in other sciences, how the Euclidean Geometry has been postulated so precisely.

Euclidean Geometry that is based on a few elementary and simple principles, after more than 2000 years of invention, still is valid in a branch of mathematics that is called plane or Euclidean geometry. This has amazed and excited many minds in the history. For example, Albert Einstein in his biography says that when he was a young school student, by studying the Euclid's Elements book has been amazed and excited from the simplicity and strength of it.[7]

Of course, most of the results of the principles that Euclid has gathered in his book, also were noted by some others before his work. But the Euclid's great work was organizing some basic simple and obvious principles to build a huge and strong

---

[7] Various Things About Albert Einstein,
http://www.einstein-website.de/z_information/variousthings.html.

logical framework that is still valid today. While many other frameworks in other fields, even formalized long after that, collapsed later. The reason for the persistency of Euclidean Geometry may be its simplicity, few basic principles, logical strength, extent of coverage and its consistency with the reality.

Also it must be added that indeed the new Non Euclidean Geometries are extensions of the Euclidean Geometry from a flat to a non flat universe and the Einstein's theory of relativity is a consequence of it. Non Euclidean Geometries have their origins from the doubts in possibility of drawing only one line from a point parallel to another line or the equality of the sum of angles in a triangle with 180 degrees, in a non flat universe.

Albert Einstein, in a speech about the "science, philosophy and religion", has said that the goal of all the sciences is to interpret or manage a vast number of facts by a least number of principles.[8] Also Galileo Galilei in comparing the Ptolemaic model (earth in center) with Copernican model (sun in center), has noted that the essence of nature is not to complicate the things unnecessarily, but is to produce more effects by simplest and easiest ways.[9]

Indeed, like in mathematics, when there is more abstraction, there is more coverage of the related basic principles. But, creating a logical framework with a few basic, simple and strong principles and a level of abstraction with a larger extent of coverage may be very difficult in practice. In science, less or more, always there have been tendency to simplification, consistency with the reality and extending the coverage regardless of its success or fail.

Here, we can note to some familiar historical cases related to the subject of our discussion. Roman number displaying method and decimal numbering method in arithmetic and Ptolemaic and Copernican models in astronomy are interesting examples in this case. The major difference between the Roman and decimal methods in numbering is the lack of importance of the positions of numerals in the Roman method and importance of the positions of digits in the decimal method. Also, the major difference between the Ptolemaic and Copernican models is earth centrism in Ptolemaic and sun centrism in Copernican model.

Apparently, a little difference in building the primary frameworks, has led to huge differences and complexities in the related methods or models. Displaying numbers in the Roman method instead of the decimal method and the interpretation of celestial events in Ptolemaic model instead of the Copernican model are very difficult and complicated.

In both the Roman method and Ptolemaic model, in dealing with some new situations, always they had to modify the old principles or to add new ones. While in decimal method and Copernican model, the simple primary principles were enough to deal with new situations. This has led to simplicity, consistency or

---

[8] Science, Philosophy and Religion, A Symposium, 1941, New York.
[9] Galileo, G. Dialogue Concerning the Two Chief World System, Translated by Drake ,1962, Berkeley.

compatibility with the reality and still remaining in decimal method and Copernican model against the Roman method and Ptolemaic model.

In Roman method, writing big numbers or doing the basic mathematical operations (addition, subtraction, multiplication and division) was very hard or impossible. While in decimal method that we learn it in elementary school and use it in our every day life, there is not any restriction in writing even many big numbers or doing very complicated mathematical calculations. Also in Ptolemaic model interpretation of the motions of other planets was difficult or impossible. But in Copernican model not only the motion of earth around the sun, but also the motion of the other planets around it are interpreted very easily.

Today, decimal method in numbering or Copernican model in astronomy is so simple, obvious and consistent with the reality for us, so that we wonder how they were not found by any one before that. Of course, as mentioned above, most of the sense of simplicity is due to their consistency or compatibility with the reality. We have ten fingers in our hands and our fingers have been the first available numbering and calculating tool for human kind. Also the reality is the motion of earth around the sun as in Copernican model and not the motion of sun around the earth as in Ptolemaic model.

At the end of this introduction, it may be interesting to say that Euclidean Geometry, chronologies, Roman and decimal methods and Ptolemaic and Copernican models regardless of the truth or falsity and simplicity or complexity, all are as a thing that we call it "system" today. One side of the concept of "system" is our imaginations about the objects and their related events in the world around, that may remain valid or become invalid by time. In this book, this view to the objects and events in the world around us is presented.

0

Whole Review of the Book

Introduction

"System" is a popular word that is used in all fields of the human knowledge for many objective or subjective things in most languages. According to the "Oxford Dictionary", "system" is among the 1000 most popular words in English language. Many people, with or without an exact purpose of this word in mind, use it more and more in their daily language in some way. This shows that the word "system" has a vast and broad use in natural languages. The importance of this usage is that "system" carries a "concept" with itself than a device name like "computer".

Also "systems thinking" as a method of thinking based on the system and its related concepts, is used in many public or scientific fields. Like "system", also "systems thinking" has a vast and broad use in most of the fields of human knowledge. To introduce the basic concepts of system and systems thinking, we start from the ordinary related concepts or basic things in the world around.

Objects and Events

Many evidences show that we are living in a world mostly consisted of the "objects" and their related "events". Objects are the things that exist in "space" in some type. Events are the actions in relation with the objects that occur in "time" in some way. Objects are the "constant" aspects (in the dimensions of space) and events are the "variable" aspects (in the flow of time) of the world we live in.

This is why that news as daily "hot" subjects and ordinary language sentences as a tool to explain the world in our every day speaking, all are about the objects and their related events! For example consider the following sentences in news or our every day speaking: "plane landed", "I am reading this book" and many other similar sentences. "Plane" and "book" are as objects that occupy "space" and "landing" and "reading" are as events related with the objects that occur in "time".

Objects and events are mutually related with each other. In other words, usually without objects no event is meaningful to be occurred. Also if we consider that every object in the world has come into existence in some event in some time, then without events no object can be imagined to be existed.

From our view in a study, objects usually are consisted of some "related parts" with their own "order" as a "whole" in the space. Also events usually are consisted of some "stepwise actions" with their own "rule" as a "happening" in time. According to this, components forming the objects have their own space-based orders and steps forming the events have their own time-based rules as the total arrangement of the objects and their related events in the space and time.

In most studies, objects from the view of their "constant" aspects, usually are not the cause of the problems. But when an object of interest is the subject of some internal or external related events, then there may be problems in relation with its "variable" aspects. In other words, in most studies, objects usually are important due to their related events! Indeed, objects without related events usually are not the subject of interest or have not role in most of the problematic cases in our world. We can note some facts to confirm what told in the above.

For example, we can start from the root and the universe itself. It can be said that based on the knowledge today and "big bang" theory, even the universe itself has come into existence form a "primary condensed matter" as an object in the primary space and a "primary great explosion" as an event in the primary time. Big bang theory tries to explain the properties of this initial object and method of this initial event. Based on this theory, even today, universe in the form of "matter" in it as a giant object in the dimensions of space, by the "energy" resulted from this initial great explosion, is expanding as a huge event in the flow of time. Note the role of "matter" and "energy" in this case. We will return back to this.

Also in the natural languages as one of the first and most common and available tools for explaining the world, "noun" as one of the most important language elements, points to "object" and "verb" as another important language element, points to "event". Without the nouns and verbs, the language has no functionality in its totality. Like the objects and events in the ordinary world, nouns are the things and verbs are the actions that occur in relation with the nouns.

Just as the objects, with nouns usually we have no problem. But when nouns are the subject of some internal or external related verbs, then there may be problems. Also like the objects and their related events, nouns without related verbs, usually have not role in most of the problematic cases. In other words, sentences with nouns and verbs usually are important due to the verb of the sentence! Indeed, sentences without verbs are not the subject of interest or even are unable to transfer a concept.

In the sentences in natural languages, nouns can have "adjective" to explain certain property or "order" of the object related with the noun. Also verbs can have "adverb" to explain certain method or "rule" of the verb related with the event that is in relation with the object. Adjective of a noun some how points to the "order" in the noun that points to the object. Also adverb of a verb some how points to the "rule" in the verb that points to the event.

Now we can summarize that in natural languages noun points to object, verb points to event, adjective points to the order governing the object and adverb points to the rule governing the event. If in the ordinary world objects are related with the events, in natural languages nouns are related with the verbs. If an object without a related event has no challenge, noun without a related verb also has no challenge. Indeed, nouns, verbs, adjectives and adverbs usually are the main words that most of the cases can be expressed with, in most of the natural languages.

This is why the first words we learn at school are the nouns that point to objects, verbs that point to events and adjectives or adverbs that express the "how is the object or event". Nouns that point to the objects imply space but not time. Verbs that point to the events imply time but not space. Also adjectives and adverbs expressing the nouns and verbs somehow imply space and time.

Structure, Behavior and Discipline
We saw that from our view in a study, objects usually are consisted of their own "ordered related parts". Based on this, it can be said that objects have their own "structure" in the space. Structure (that usually is "matter" based), is the base of existence of the objects and implies some aspects of "constancy" or "silence". For example, consider the "structure" of a "plane" or a "book".

Also, objects due to their usually "stepwise related events", have their own "behavior" in the time. Behavior in general (that usually is "energy" based), is to do some thing or to have some application or functionality. This is done by the aid of an internal or external source of "energy". Indeed, behavior of a thing is a general flow of getting "inputs", to "function" on it in order to produce "outputs" as the result of functioning. Behavior implies some aspects of "variation" or "motion". For example, "reading" is the "application" of "book" and "landing" is part of the "functioning" of "plane".

According to the above, structure is governed by its own "structural order" and behavior by its own "behavioral rule". Now we can say that objects due to the order in their structure and rule in their behavior, have their own "discipline" in the space and time. In other words, discipline is the order governing the structure and rule governing the behavior of the object due to its related events. Indeed, discipline in the structure and behavior, implies some aspects of "constancy or silence" and "variation or motion" in the "space and time".

Objects that usually are "matter" based, experience constancy or silence and variation or motion by the aid of some external or internal agent that usually is "energy" based. Discipline, as we will see, is "matter", "energy" and "information" based and causes the objects always somehow "to protect" their structural order and "to control" their behavioral rule. In other words, discipline causes the objects to be existed and to continue their existence as they are. Without structure, no behavior is possible. Also, structure and behavior are possible under a certain discipline.

Now we can summarize that from our view "objects have their own structure, behavior and discipline" and based on these three, have their own "wholeness". With this view, structure is space based, behavior is time based and discipline is space and time based. In the daily languages, with a "noun", we usually remember the image, shape or "structure" of the object. With a "verb", we remember the application, function or "behavior". "Adjective" or "adverb" reminds us the "order" or "rule" in the object or in its related event. In other words, we can understand "discipline" in the structure and behavior of the object by the adjective or adverb.

For example, based on the big bang theory, "structure" of universe includes the "matter" existing in. "Structural order" of universe includes the arrangement of the matter in the space in the form of stars and galaxies. Also, "behavior" of the universe includes expanding by the "energy" resulted from the initial big bang. "Behavioral rule" of universe includes the method of expansion in all sides in the time.

In the case of universe, "discipline" is in the form of "orders and rules governing the space and time and any thing in it". In its broad concept, discipline of the universe can be titled as the "laws governing the structure and behavior of the universe". In other words, in the world we know, discipline is the orders and rules in the objects and their related events. From this view, the aim of the sciences is to find the disciplines or orders and rules in the objects and their elated events in different fields of human knowledge.

In the history, man first has studied the objects and their governing orders and then the events related with the objects and their governing rules. Oldness of "geometry" respect to "algebra" might to confirm this. Indeed, in the world around, geometry was related with the constant aspects or the structure and algebra with the variable aspects or behavior of the objects.

Man has used arithmetic to move from the "quantity" of the constant structural properties of the geometric objects in the space to the "quality" of the variable behavioral methods of the algebraic events related with, in the time. For man, in most cases not the objects with their constant aspects, but their related events with their variable aspects ("objects with their related events"), were as challenging things.

History also reveals that man has been more successful in dealing with the objects than the events. Probably, the long lasting validity of an old science like Euclidean geometry that was related with the constant aspects of the objects was due to the less challenging aspects of the constancy. In contrast, invalidity of astrology as a method of thinking about the world that was related with the variable aspects of the objects, was due to more challenging aspects of variation.

Matter, Energy and Information

As mentioned above, normally "structures" are possible by the aid of "matter" and "behaviors" by "energy". In other words, matter is the maker of structures and energy is the driver of behaviors. In the solid things in nature, discipline is deterministic and usually is possible by the thing's structural order due to matter or behavioral rule due to energy. This order or rule may be possible due to the "natural strength" (from structure or matter view) or being in a "field" or "environment" (from behavior or energy view). Indeed, in this case "discipline" is "matter" and\or "energy" based.

But in the living beings that besides the matter and energy can use "information", discipline is partly non-deterministic and information is the agent of the non-deterministic part of discipline. From this view, information as the maintainer of disciplines is in the same level as matter and energy. Indeed, in this case "discipline" is "matter" and\or "energy" and\or "information" based. Things according to their type, for existing or to continue their existence, are based on or need these three main and basic resources of matter, energy and information.

Matter is related with "mass" and energy with "force". In physics, mass or matter generates gravitational field and its resulted gravitational force or energy. Mass is the agent of structure in the space. Force is the agent of behavior in the time. Also gravitational field due to mass or force that affects the space and time is the agent of discipline in the space and time. In consequence, discipline causes "protecting the structure" and "controlling the behavior" of the celestial objects like the moon and its rotation around the earth or the earth and its rotation around the sun.

Information is the reflector of constancy and variation in the objects and their related events and finds its primary concept from this point. Indeed, information originates from the changes in the structure due to matter and especially behavior due to energy. From the view of this book, "information" is what that is related with the "organization of matter" in the "structures" in space and "management of energy" in the "behaviors" in time. By adding information to matter and energy, more complex forms of discipline in the things that can use it become possible.

As a result of above, structure and its order and behavior and its rule are some how related with the concept of "organization". Also protecting structural order and controlling behavioral rule are some how related with the concept of "management". This is why both of these two (as discipline) are important subjects in the contemporary era as the "information age".

Indeed, information as the reflector of constancy and variation is the descriptor of structure due to matter and behavior due to energy. Information is related with the usually constant structural properties and usually variable behavioral methods. So describes the related case in some way. By the aid of information, in addition to objective things, subjective things also can be explained and existed in some way.

Man first has studied the matter as the maker of the objects or structures and then the energy as the driver of the events or behaviors related with the objects. Here, oldness of "chemistry" respect to "physics" might to confirm this. Indeed, the subject of chemistry was related with matter as the constant aspects or structure of the objects. Also the subject of physics was related with energy as the variable aspects or behavior of the events related with the objects.

The history of ancient philosophy is full of battles about the constancy (structure) and variation (behavior) and the priority of one to other. For example, we can see the effect of this in the core of the Marxism theory. But for us today, like the "objects" and "events", "constancy" and "variation" are mutually related with each other. Without the one the other has not any meaning. Also according to the modern science, "chemistry" and "physics" or "matter" and "energy" are related with each other.

The above discussion about the sciences can help us to find that why the mathematics (especially geometry and algebra), chemistry and physics are as the "basic sciences". This also helps us to say that why besides the chemistry (for matter) and physics (for energy), we need a new basic science. This new science may be "cybernetics", "informatics" or "organization and management science" (for information) in order to deal with the disciplinal aspects of the things in the world around us.

Also from the view of the subjects discussed above, man has tried many ages while living on the earth. The ages also can be classified in three main eras related with the three main and basic resources of matter, energy and information. In other words, from a general view, struggles of human kind on the planet have continued by using different dominant resources in different times as follows:
- Matter age: Pre-industrial era, matter – driven age like "stone age".
- Energy age: Industrial era, energy – driven age like "steam age".
- Information age: Post-industrial or contemporary era, information – driven age like "communication age".

What is System?
In the English language dictionaries, various roots for the word "system" can be found. For example, in the internet versions of "American Heritage Dictionary of the English Language" (http://ahdictionary.com/) and "Oxford Dictionaries" (http://oxforddictionaries.com/), we can find "systema" from Latin and "sustema" from Greek and also "synhistanai". "Systema" means "to be together" and "sustema" means "to put together". In "synhistanai", the "syn" like the "co" in the beginning of the words, means "together" and "histanai" means "setting up". So, "synhistanai" means "setting up together". According to the many interpretations about system, system is a set of related things with special arrangement acting in a special way.

Based on the above and knowledge of today, "system" in its most general and complete concept, is any "objective" or "subjective" thing consisted of its "related parts" that has its own "structure", "behavior" and "discipline" as a "whole". Discipline of the system is in the form of its "structural order" and "behavioral rule". Systems usually have their own certain "application" (usually by structure only), "functionality" (usually by structure and behavior with deterministic discipline) or "goal to reach" (usually by structure and behavior with non-deterministic discipline). The wholeness of a system is greater than the sum of its parts. But, in practice, based on the conditions, our purpose of system might to be structure, behavior or discipline or a combination of these three.

The combination might to be: structure + behavior, structure + discipline (structural order aspect of discipline), behavior + discipline (behavioral rule aspect of discipline) or structure + behavior + discipline (structural order and behavioral rule aspects of discipline). In other words, all of these cases can be explained under the general title of system. This is why the meaning of system in most dictionaries is also the synonym of structure, behavior, discipline (order and rule) or a combination of these basic concepts.

General System Types
According to the above, we can define "system" in its simplest form as "a set of certain objective or subjective related elements in the form of a whole thing in general with a certain structure and structural order". In this case, things as "structures" due to matter (like Eiffel Tower in Paris) usually have their own "application". Things of this kind are subject of interest for their researchers from the view of their "structure". Application of the Eiffel Tower as a "structure" was due to its symbolic aspect in a fare when it was established and then as a tourist attraction place. Here, we are faced with constancy and no variation. This kind of systems mostly also can be called as "static system".

We saw that objects usually are the subject of some related events and due to the events, have some kind of behavior. Hence, objects and related events must be considered together. In other words, objects with their own structure, due to the events occur for them, have their own application or functionality as behavior. Here we can complete the previous concept of system (as a structure only). Now we can say that "system is a structure with a certain structural order that due the events occur for it, has its own application or functionality as behavior with a certain rule".

In this case, we have things as structure due to matter plus application or functionality as behavior due to energy. Things of this kind (like Big Ben clock in London) are subject of interest for their researchers from the view of their "structure and behavior" and especially "behavior". A "structure" with a certain "behavior" usually is considered as what we call it "mechanism". Here, we are faced with deterministic or non-information based discipline. This kind of systems, mostly also can be called as "dynamic system".

Structural order and protecting it and behavioral rule and controlling it, represent the governing "discipline" in the systems. When there is talking about the behavior and discipline, inevitably there is talking about the "application", "functionality" or "goal to reach" in the behaving structure. Because, protecting and controlling the discipline in the structure and behavior, usually is meaningful when there is a specific application, functionality or a goal as a destination to reach.

With this interpretation, survival of systems in order to have their own application or functionality or to achieve their own goals needs the survival of structure with its order in the space and behavior with its rule in the time. Having or achieving this becomes possible by "structure and structural order protection and behavior and behavioral rule control mechanism" in systems.

From a general view, application, functionality or goal in systems are meaningful in relation with behavior. In other words, any system during its usage or life time, can have certain application or functionality or to seek certain goal. Dynamic systems acting in this way, in its general concept can be called as "application or functional systems" or "goal seeking systems". Application or functionality usually is possible by the aid of an external (or internal) source of energy. But goal seeking usually needs internal (or external) source of energy. This passes us from deterministic or non-information based to non-deterministic or information based discipline.

We came to "static system" by starting from the object with a certain structure only and then came to "dynamic system" by adding behavior to this structure. In this level, we had deterministic or non-information based discipline. Now we add non-deterministic or information based discipline to systems. Here, again we can complete the previous concept of the system. In this level, "system is a certain ordered structure with a ruled behavior that in the form of its governing discipline as a whole, by protecting its structural order and controlling its behavioral rule, has certain application or functionality or seeks to achieve certain goal".

Structure protection and behavior control with the aid of information, when is done internally and by the system itself, is a sign of "being alive" and can be seen in the living beings or "organisms". Besides the static and dynamic systems, this kind of system mostly also can be called as "organic system". This completes the three general types of systems based on our discussions.

Systems Protection and Control Mechanism
As we saw, any system in order to remain as it is, in a least condition, has its own "structure and structural order protection mechanism". For example, a piece of stone if wants to remain as a piece of stone as it is, naturally resists against the breaking and collapsing. In this example, the natural strength of the system has the duty of protection.

Also, any system in order to continue its behavior, in addition to "structure and structural order protection mechanism" has its own "behavior and behavioral rule control mechanism". For example, an ordinary door if wants to remain as a door as it is, while naturally resists against breaking and collapsing, also resists against the unusual movements out of what the hinges make it possible. In this example, the natural strength or what the maker of the system has fixed in it has the duty of protection and control.

In old artificial systems like ordinary doors, behavior is possible by the aid of an external source of energy, for example the energy or force of the passer of the door. But in the new artificial systems like electronic doors, the source of energy is internal and by the system itself. In both of the old ordinary and new electronic doors, the "protection" duty is the same. In other words, it is possible by the aid of natural strength or what the maker of the door has fixed in it. But "control" duty in ordinary doors is external and in electronic doors is internal.

Organic systems like living beings have their own "structure and structural order protection and behavior and behavioral rule control mechanism". Some part of this "protection and control" duty is internal and some other is external. In other words, in organic systems some part of the protection and control duty can be internal and to be done by the system itself.

Stability in Systems
According to the above, in systems, protecting the structural order and controlling the behavioral rule is possible by its external or internal "structure protection and behavior control mechanism". This leads to some kind of "stability" that we can see in the objects with their related events as systems in the world around. In other words, while the real world is accompanied by constancy and variation, but naturally, systems tend to have certain and fixed properties or attributes and to be in a desirable and persistent situation or state between constancy and variation that is called "stable state".

Stability is possible when the system has balance in its structure, equilibrium in its behavior and certainty in its discipline. These three in sum bring the total stability for the system. So, we can say that total stability in systems is the resultant of their "structural balance" due to matter, "behavioral equilibrium" due to energy and "disciplinal certainty" due to information.

Entropy in Systems
"Entropy" as an "inverse scale to display the presence or absence of energy for doing work", first derived from the second law of thermodynamics. The laws of thermodynamics are related with energy in general as the cornerstone of industrial revolution. So, from energy view, when there is energy to do work, we have less entropy and when energy ends, entropy increases.

Also later, the concept of "entropy" was used in information theory. In information theory, information is defined as the "order" against "disorder" or "certainty" against "uncertainty". Considering "order" and "disorder" (and adding "rule" and "misrule" to this) as a scale of having or not having information, connects entropy concept from energy view to its concept from information view.

In other words, when there is energy, the system can go in its right way and we have the desired order and rule governing the system. Indeed, in this case we have the desired structural order and behavioral rule in the system from matter and energy view. But when energy ends, disorder and misrule replace the order and rule in the system. If information also can be used in system, having the desired information leads to "certainty" and not having it leads to "uncertainty" in the "discipline" of the system from information view.

As we see, the concept of entropy in systems is very comprehensive. A comprehensive definition of entropy can include the three main and basic resources of matter, energy and information and also the structure, behavior and discipline. Based on the above, the concept of entropy can be extended to be as the scale of "order" and "disorder" in the structures, "rule" and "misrule" in the behaviors and "certainty" and "uncertainty" in the disciplines.

In other words, entropy can be considered as a factor of presence or absence of order in the structures (matter), rule in the behaviors (energy) and certainty in the disciplines (information). Entropy also can be viewed as a criterion or scale for "weakness or strength" of relations among the parts from structural view, "stasis or dynamism" from behavioral view and "regularity or irregularity" from disciplinal view. From this view, we can evaluate entropy in all systems.

In a normal condition, entropy is like "time" and as a total scale for "lifetime" or "age" tends to increase by time. Consider a system "is left to be as it was" or with no inputs in the form of matter, energy and information. This system usually tends towards "weakness or losing the relations" among the parts from structural view, "stasis" from behavioral view and "irregularity" from disciplinal view. In consequence, the entropy of it increases. But if the system "is not left to be as it was", by getting the required inputs and interacting with the environment can slow down the normal increase of entropy.

Indeed, there is no system to remain in the same condition or state for ever. Evidences show that with the unidirectional increase of time, in a natural trend, all systems objectively or subjectively tend towards some kind of structural destruction and extinction, behavioral stasis and silence and disciplinal disorder and misrule in general. Finally the system may change its state or reach to an end as it was.

One of the ways to overcome the increase of entropy in the systems is to apply some kind of "maintenance". Here, "maintenance" means continuously supplying of required matter for keeping the structure to be fixed, energy for keeping the behavior to be done and information for keeping the discipline to be updated in the case of structure, behavior and environment.

For example, consider a static system like an ordinary building. In this system only the structural aspect of the system is important. In this case, "maintenance" is continuously strengthening of the structural relations of the structure in order to keep it as it is. This is exactly the work that is done continuously on the structures like Eiffel Tower in Paris.

In the case of dynamic systems, like the static systems, energy must be input to the system in order the system to continue its behavior. This is exactly the work that is done continuously on the mechanisms like Big Ben clock in London. Also for example, in order a coal energy driven train to continue its behavior, coal as the desired energy resource must be fed into the fire place of the train continuously, as we see in the films with old stories.

This is also true for the systems that need to use information. In other words, all the systems that need to use information in order to do their duties, need to receive their required information continuously. In transcendental organic systems like human beings, need to receive information is even more vital than matter and energy. Because transcendental organic systems usually have internal reservoirs of matter and energy, but need to receive the desired on-the-time information just at the time that the system needs it during behaving.

Systems Life Cycle
To continue having structure and protecting the structural order and having behavior and controlling the behavioral rule in space and time, can be viewed as the "system life". By this view, any system as its "life cycle" can "be structured or born" in one time, "behave or live" for a length of time and finally "go out or die" in another time. This may be viewed once for ever or in a repetitive form. From this view, we can evaluate different kinds of life cycle for systems. System life cycles can be in two major types as "linear\ start – stop life cycle" and "circular\ periodic life cycle". These two life cycles can be viewed also as "open" and "closed" life cycles.

Linear or open life cycle is a linear or open path consisting of "start – exist – stop" or like in living beings "birth – live – death". In circular or closed life cycle this path is repeated periodically. The repetition process like "turning on - operation - turning off" in an ordinary electrical device, may be the same in different repetitions or like in plants may be different. In other words, the start and stop is repeated along the time with no change or change. From systems view, in the case of systems life cycles, instead of a linear or "open" life cycle, usually we are faced with a circular or "closed" life cycle.

Based on the evidences, in a normal condition, the endless flow of time in systems leads to "growth" and "evolution" in one side and "destruction" and "extinction" in the other side. Going of "old" and coming of "new" is the manifestation of this trend. In other words, "dynamism" in general that includes "behavior", with its positive or negative dimensions, can be imagined in three main forms of "growth", "destruction" or "oscillation form of growth and destruction" as a combination of these two.

Usually, "growth" can lead to "evolution" and "destruction" to "extinction". Evidences also show that the behavior of most of the dynamic systems some how follows these three basic and main patterns. These patterns also can be imagined in different forms. Indeed, based on these basic patterns in the behavior of systems, we can imagine different forms of behavior in systems. The forms differ from uniform linear to non-linear growth or destruction. But with a delay, the direction may change to opposite side. In other words, growth may change to destruction or destruction to growth. This may continue in a wave or sinusoid form. This is why we use lines and curves in graphs to represent the behaviors.

Determination and Free will
As we saw in the above, discipline can be discussed in two main categories of deterministic and non-deterministic disciplines. In deterministic discipline there is not any intelligent being or device intervention that can use information in organizing matter and managing energy. But in non-deterministic discipline usually there is some kind of this intervention.

Deterministic discipline is what we see in the form of natural organization and management of matter and energy in the nature. For example, consider our planet with mountains and valleys as the organization of matter on it. Also remember what we told about the moon and its rotation around the earth or earth and its rotation around the sun as the organization of matter and management of energy in them.

Now consider for example the early man made water mills. In the early water mills, with the high or low energy of the flowing water, there was high or low speed of rotation in the upper stone of the mill. By a simple feed-back mechanism in the structure of the mill, the slope of the grain feeder to the mill was changed by the change in speed of rotation of the upper stone. This "managed" the amount of grain entered to the mill by the change in the energy of the flowing water. In other words, with high or low energy of water (high or low rotation of upper stone), more or less grain entered to the mill with no human interventions.

Devices like early water mills are the first cases of man made devices that some how use information in the form of simple feed-back mechanism. In this case, information is related with the intensity of flowing water. This simple feed-back mechanism manages the behavior of the mill without any human intervention. In other words, this kind of devices had some kind of a simple and primary intelligence.

In the nature, there is a spectrum of the things in the form of solids, plants and animals. At the one end of this spectrum and in the case of solids, usually there is a matter or energy based bed (some thing like field in physics or environment in nature). This bed encompasses the thing and forces it to have a usually constant structural order and behavioral rule.

Solids have structure and structural order. Plants in addition to structure and structural order have some kind of behavior and behavioral rule. Living beings in addition to structure and structural order and behavior and behavioral rule, have their own mechanism for protecting and controlling discipline governing them.

For example, in the case of solids, a piece of stone silent in its place in the nature has its own constant structural order. The piece of stone has no authority to do any thing for example to change its place. In other words, by ignoring the internal, gradual and natural changes of the stone, all the changes are from an external source. Also moon is in the field of energy or force of the gravity due to the matter or mass of the earth. Because of this, moon has its own constant structural order (as a sphere) and behavioral rule (as rotating around the earth).

Plants have their own rule of growing or repetition of being green in the spring and yellow in the fall. From this view, plants have a kind of closed circular life cycle with progressive growth or destruction during their life time. Plants also have the ability of using a level of information related to their environment. For example, they can follow the sun light. But they are unable to furnish their environment. For example they can not change their place or bringing water to their root, if they face with dry years. In the case of earth and its rotation around the sun, because of its biosphere or what that is called "noosphere" (the sphere of "information"), the condition of repetition is more complex.

At the other end of this spectrum and in the case of animals and especially human beings, some part of discipline is possible by interacting with the environment and by the aid of information. This is done while there is dependency to the environment also. Living beings while can use minor resources of energy when compared with huge forces ruling the nature, by the aid of information can improve their abilities in dealing with the environment they live in. This is why that information is power!

Discussing about the discipline, according to the ability or inability of using information is related with determination and free will or non-freedom and freedom. In other words, the order and rule due to the discipline in the structure and behavior, in different cases has "non-freedom", both "non-freedom and freedom" and "freedom" aspect.

In non-free disciplines, the external protecting aspects and in free disciplines, the internal control aspects are dominated. In non-free disciplines, due to the governance of matter or energy (mass or force), there is some thing like a field or environment that encompasses the objects and there is no freedom. But in free disciplines, due to the role of information, objects have some kind of freedom.

In this spectrum, from solids to animals and especially human beings, determination or non-freedom aspects in discipline decrease and non-determination or freedom aspects increase. Information plays an important role in this decrease and increase. It means that in the cases that besides the matter and energy, also information can be used, there is more freedom. But if in the one end of this spectrum there may be absolute determination, in the other end of it there is a relative non-determination or authority.

For example, a piece of stone as an object with a structure, in most cases is important due to its material constant and silent aspect as a piece of stone. In this case, structural order of the piece of stone deterministically is the same as it was in the beginning. If the piece of stone is inside flowing water that can move it, then in addition to structure due to matter, will have a behavior due to energy of the flowing water. In the above example, in the first case we are faced with structure and structural order due to matter and in the second case in addition to this, with behavior and behavioral rule due to energy.

Indeed, in the first case the piece of stone while internally has no freedom, but externally is free and its external freedom is a result of its internal non-freedom. In the second case the piece of stone is in a field or environment that the flowing water represents it and dictates the behavior of the stone to it. What told about the behavior of the piece of stone also can be true about the moon and the field of gravity of earth. In this example, piece of stone or moon have no freedom and what forces them to behave as they behave, is the field or environment due to the force or energy of water flow or gravity of earth.

Now instead of stone and flowing water, consider a boat with a boatman in a river or sea. In this case, the set of boat and boatman while are in a deterministic field or environment like before (with force or energy due to the flow of water or wind), will have some degree of freedom. This freedom results from the ability of getting and processing the information. For example, boatman gets the desired information from the position of stars in the sky or a compass and processes it. Boatman then uses the energy of arms to run the boat in the right way.

Now instead of boat and boatman in the river or sea, consider a spaceship and astronaut in the orbit of the earth. Then more or less what told about the boat and boatman also will be true about the spaceship and astronaut. In this case, spaceship and astronaut while are in a field of force or energy, by the aid of information and the energy saved in the spaceship, can select the right way. But the moon with the similar position in the space has not such an authority.

Now we can understand why in the modern world of today there is so much sensitivity and discussions about the free flow of information in the human societies. Discussions are also about the free exchange of material products and free trading of it and energy generation and transfer in the form of electricity for all humans. Indeed matter, energy and information are the three main and basic resources needed in systems.

What is Systems Thinking?

System and its related concepts is a powerful tool to deal with any objective or subjective thing as a whole. In dealing with the things, its parts and relations among them, its function and steps of it and the order and rule governing the thing as a whole, are considered. This approach includes seeing the outward (exterior) or structure, inward (interior) or behavior and the structural order and behavioral rule or discipline governing the thing as a whole.

This is done under the concept of system and systems thinking. System and systems thinking include a set of concepts and methods in the case of objects and their related events and their governing orders and rules. From this view, things have their own structure, behavior and discipline as a whole.

In other words, system and systems thinking are the general title of the concepts and methods that are used to explain approximately any thing in the world. This type of approach to the world is common and natural. As we saw in the above, the trace and effect of it can be seen in the natural languages as one the most common and naturally available tools to explain the world by man.

Indeed, system and systems thinking are the combination of the two old trends in dealing with the world. These trends are "reductionism" and "holism". Reductionism emphasizes on the parts forming the objects. Holism emphasizes on the whole due to the relations among the parts forming the object in order to have an application or functionality or to present a behavior to achieve a goal. Reductionism by emphasizing on structural elements led to some kind of "structuralism". Holism by emphasizing on the behavioral wholeness, led to some kind of "behaviorism".

Systemic, Systematic and Systems Approach
Based on the above, reductionism can be called as "systemic approach", "structure and structural order based approach" or "order based approach". Holism can be called as "systematic approach", "behavior and behavioral rule based approach" or "rule based approach". Daily uses of the words "systemic" and "systematic" or their meaning in dictionaries, some how confirms this interpretation.

Dictionary definition of "systemic" means what that are "built in the structure" of the system. Also "systematic" usually means "doing some thing by using a ruled or stepwise method". The combination of these two approaches also can be called as "systems approach", "systems thinking", "structure and structural order and behavior and behavioral rule based approach" or "order and rule based approach".

"Systems thinking" also may be introduced form "organization" and "management" view. "Organization" is related with structure and structural order and behavior and behavioral rule. "Management" is related with structure and structural order protection and behavior and behavioral rule control. This is why system and systems thinking are the subjects of interest in organization and management sciences.

A General System and Systems Thinking Approach

Language is as the "tool of tools" in every thinking activity, including systems thinking. One of the practical methods of using the ordinary language for explaining systems is so-called "brain storming" method. This is done by thinking about the subject of interest and writing any thing coming to mind about it in the form of ordinary sentences even without any organization.

This writing, after initial preparation can be reviewed again and again in different times and the new things coming to mind, added to it. The first step in this work is to be certain that approximately "all the things" known about the subject of interest, are written. This writing is the base of the work and can be titled as the raw "system statement". Now we can extract the "objects and their structural orders", "events and their behavioral rules" and "related objects and events" from this statement.

To start, first we can extract all the nouns or noun implied words from the "system statement" with ignoring the cases that are not related with the subject of interest. Then, among the extracted and refined nouns or noun implying words, we can determine common and proper nouns and write each proper noun in ordinary face under its related common noun that is written in bold face. This is done for all the common nouns and related proper nouns until all the extracted nouns are written in a column. Then we can find the adjectives of the nouns in the system statement and write it in front of its related noun in the extracted list of nouns. This is done until all the extracted adjectives are written in rows in front of the nouns.

The same work must be done in the case of compound verbs and the related simple verbs under each and adverbs of verbs related with the verbs. In other words, like nouns and adjectives, compound verbs are written in bold face and related simple verbs in common face under each in a column and the related adverbs in the rows in front of the verbs.

Now, in the first list, in the case of nouns determine if the noun is subject (actor of the verb) or object (verb done on it). Display this by writing "S" (for subjects) or "O" (for objects) in the right-down corner of the related noun. Also in the second list, in the case of verbs determine the time of the verb. Like before, display this by writing "Ps" (for past), "Pr" (for present) and "Fu" (for future) in the right-down corner of the related verb.

For completing the work, combine the first and second lists by intersecting them. Write the result in a third list as a table. In other words, in the table assign the rows to nouns and columns to verbs. To do this, copy all the verbs to the first row and nouns to the first column of the table for intersecting them. Now by cross referencing the nouns and verbs in the rows and columns of the table, determine if a noun is related with a verb and put an "X" in the intersection of the related row and column.

Now it can be said that from the written "system statement", a raw list of objects (nouns), related events (verbs) and its orders and rules (adjectives related with nouns as structural orders and adverbs related with verbs as behavioral rules) are extracted.

In the extracted cases, approximately each common noun can be viewed as a "group of objects" or system. Also, approximately each proper noun under a common noun can be viewed as an "object" or an element of a system. Like this, approximately each compound verb can be viewed as an "application", "function" or "behavior" in a system. Also approximately each simple verb under a compound verb can be viewed as one aspect of the behavior of the related system.

To find the cause and effect relations or processes, in the third list, especially the subjects (an aspect of cause), verbs (as event due to the cause) and objects (an aspect of effect) can be helpful. Also, approximately it can be said that the "history of the system behavior" is extractable from the past tense verbs, "current behavior of the system" from the present tense verbs and "forecasting the future behavior of the system" from the future tense verbs.

Now the structure and structural order, behavior and behavioral rule and the protection and control mechanism for protecting the structural order and controlling the behavioral rule or the discipline of the system also can be explained. As we see, if a certain subject of interest is explained in a "certain" way, then we can identify the related "system". This is done by writing the system statement mentioned above, then extracting the "system concepts" and finally "combining" them in order to achieve to a "whole" as a system.

By applying the structural aspects in the system concept, processed "system statement" can have a certain structure and used as system explanation. The written system statement in a well formed state can have the following structure:
- Starting material
- Body of the text
- Ending material

Final Note
Today, in many cases, "lack of system" means "lack of order in the structure, rule in the behavior and mechanism for protecting the structure and structural order and controlling the behavior and behavioral rule". Important aspect of the era we live in is paying attention to the organizational and managerial aspects of the objects and their related events. This is why the two subjects of organization and management are among the most important subjects, especially in the field of systems.

As a final word, it can be said that one side of the concept of "system" is our imaginations about the objects and their related events. This is why for the same thing but from different perspectives or views we consider different systems. This imagination may remain valid or change and become invalid by the change in view or time. In this book, this view to the objects and events in the world around is presented.

1

General Concepts

Introduction
In the "Introduction to the Book" (Chapter -1), we saw that the "system" and "systems thinking" and related concepts, not only used in the field they first introduced in, but also widely used in many other fields. Based on this, it can be said that system and systems thinking even have acted as a public and universal principle and tool. In this chapter we will see why the system and systems thinking can be viewed as a "fundamental theory and practice".

Objects and Events
Many evidences show that the world we live in is mostly consisted of "objects" and "events". Based on the science today and the "Big Bang Theory" even the universe has originated from an "initial highly compressed matter" as "object" and an "initial extra huge explosion" as "event". Objects are the "constant" aspects and events are the "variable" aspects of our world.

Object is any objective or subjective thing in general that "exists" in "space" in some type and event is any objective or subjective action in relation with the objects in general that "happens" in "time" in some way. Objects have certain properties and their related events occur according to the certain patterns. Components forming the objects have their own space-depended and steps forming the events have their own time-depended arrangement. Based on this, objects have their own "order" and their related events have their own "rule" in "space and time".

In natural languages as the first and main tool to explain the world, "noun" is used for the "object", "verb" is used for the "event", "adjective" is used for the "how of the object" or order in the objects and "adverb" is used for the "how of the event" or rule in the events. If the world is mainly consisted of objects and events, then the sentences in natural languages also are consisted of nouns and verbs. If we put a side the nouns and verbs from the language or sentences, then language becomes useless in its totality and sentence unable to transfer the concept. This confirms the main role of objects and events in the world we are faced with.

Nouns indicating the objects imply "silence" or "constancy" and verbs indicating the events imply "motion" or "variation". In the world we are faced with, objects with their own constancy usually have no challenge but events with their own variation usually are challenging cases. So is in the case of sentences in languages. In other words, nouns usually have no challenge but verbs usually are challenging cases.

This is why the first words we learn in school is "noun" that points to "object", "verb" that points to "event" and "adjective or adverb" that expresses the "how is the object or event". Also the nouns that point to the objects, imply space but not time, verbs that point to the events, imply time but not space and adjectives and adverbs expressing the nouns and verbs some how imply space and time.

Structure, Behavior and Discipline

From our view, objects usually are "a set of related parts as a whole" with their own order and some events may occur for them with their own rule. "Part" is the constituents, "relation" is the dependency among the parts and "whole" is the unity of constituents in the form of the object. Based on this, it can be said that objects have "structure" with their own "structural order" and due to the events occur for them, have "behavior" with their own "behavioral rule". If we put the structural order and behavioral rule under the general title of "discipline", then it can be said that "objects have their own structure, behavior and discipline".

For example, the "structure" of the universe mainly is in the form of celestial objects and relation among them in the form of gravitational forces. The "behavior" of the universe mainly is in the form of all-sided expansion. The structure of universe has its own "order" and the behavior of universe has its own "rule". The outcome of this order and rule is the existing state of the universe and universe due to this discipline in its structure and behavior still continues to exist.

Also, in a general speaking, it can be said that in the nature solids have structure and structural order. Plants in addition to structure and structural order have some kind of behavior and behavioral rule. Living beings in addition to structure and structural order and behavior and behavioral rule, have their own mechanism for protecting and controlling discipline governing them.

"Structure" is any objective or subjective combination of parts that are related to each other and at sum, display a certain appearance of the object in the space. "Behavior" is any objective or subjective action or reaction that an object like animal, plant or man made device does it against the internal or external events. Also "discipline" includes "structural order" as the arrangement of parts in the space that must be protected and "behavioral rule" as the arrangement of actions and reactions in the time that must be controlled. This is done by the "feed-back loops" and in the form of "structural order protection and behavioral rule control mechanism" or "protection and control mechanism" in short.

"Structure" has "static" or constant and silent aspects and implies some aspects of "matter" or "mass". "Behavior" has "dynamic" or variable and moving aspects and implies some aspects of "energy" or "force". Also "discipline" as the order in the static structure and rule in the dynamic behavior or "organization" and protecting structural order and controlling behavioral rule or "management", implies some aspects of "information". From this view, "organizing" is to generate the structure and define the behavior and "managing" is applying a mechanism to protect structure and structural order and to control behavior and behavioral rule.

Structural order is based on "balance" in the structure from "matter" or "mass" view and behavioral rule is based on "equilibrium" in the behavior from "energy" or "force" view. Like this, protecting structural order and controlling behavioral rule in the objects that can use information (and so are with some degree of freedom), is based on "certainty" or "confidence" of "protecting order" in the structure and "controlling rule" in the behavior from "information" view. In the objects that can not use information (and so are with no degree of freedom), usually there is some type of "field" or "environment" caused by or made of matter and energy that encompasses the object and like solar system as an object, forces it to have its own discipline in its own structure and behavior in the space and time.

"Balance" is a kind of equality of matter or participated mass in the structure and "equilibrium" is a kind of equality of energy or participated force in the behavior. From this view, "certainty" is a kind of equality of confidence and doubt, information or participated knowledge affecting the structure and behavior that leads to the establishment of order and protecting it and applying of rule and controlling it from organizational and managerial view. While mass implies matter and force implies energy, organizing mass and managing force implies information and knowledge.

Continuation of balance in the structure, equilibrium in the behavior and certainty in the discipline, leads to a state that is called "stability" and all the objects normally tend to be in this state. Stability in objects needs "structural balance" from matter view, "behavioral equilibrium" from energy view and "disciplinal certainty" from information view and these three are the main factors of stability in the objects. Indeed, stability is the outcome of some kind of coordination among the structure, behavior and discipline in the objects.

As we will see, this is the same concept that based on the knowledge today, can be said about "system". In other words, system is a set of objective or subjective related parts as a whole that some events may occur for it. System, due to the wholeness of its related parts has its own structure, due to the events occurs for it has its own behavior and due to the necessity of protecting its structural order and controlling its behavioral rule in order to achieve its goals, has its own discipline.

The concept of system can be viewed in three main levels as follows. In the first level: an "object with structure". In the second level: a "structure and behavior based on this structure". In the third level: an "ordered structure with a ruled behavior and structural order protection and behavioral rule control mechanism". In other words, final concept in addition to structure and behavior also includes the order in the structure and protecting it and rule in the behavior and controlling it or discipline in the object. It can be said that the third level includes the organizational and managerial dimensions in the objects and their related events that both are the challenging subjects in the contemporary world.

Matter, Energy and Information
"Matter", "energy" and "information" are three main and basic resources that we can trace their role in studying the objects and their related events. As we saw, mass needs matter, force needs energy and organizing the mass and managing the

force need information. From ancient times, "chemistry" and "physics", as two main and basic subjects in science, implicitly or explicitly have discussed the mass or "matter" and "structure" (chemistry) and the force or "energy" and "behavior" (physics). In the case of "information" and "discipline", there has been new subjects of science and technology such as "cybernetics", "informatics", "organization", "management" and "Information and Communication Technology = ICT".

These resources are needed for generating structures, performing behaviors and organizing structures and managing behaviors in the objects. In relation with an object, matter implies some kind of "constancy" in space by structure, energy implies some kind of "variation" in time by behavior and information implies some kind of "how the structure to be constant in space and how the behavior to be variable in time". In summary, we can say that matter has role in the structure, energy has role in the behavior due to the related events and information has role in protecting the order in the structure (organization) and controlling the rule in the behavior (management) of the objects.

"Matter" or "structure" implies some kind of "potentiality" or "in-active power" that any object has in its constant and silent state. "Energy" or "behavior" implies some kind of "actuality" or "active action" that any object has in its variable and moving state. It can be said that any "actuality" comes from "potentiality" and any "variation" comes from "constancy" and this generates or carries "information".

Each of these three main and basic resources some how include each other and it can be said that energy is derivable from matter and information from matter and energy. In other words, as the matter can be a resource for generating energy, also the matter and energy can be a resource for generating information. Change resulted from energy, includes information resulted from change. In the past, mainly the matter was at the center of interchanges and then energy was added to it. Now, the information is as a resource for interchange. Today, the rate of information interchange is the main factor of development in societies in a general sense.

The word "matter" is from the root "materia" with the meaning of "substance that makes objects". Matter is the structural substance of the objects that has mass and occupies space and implies some kind of constancy. This definition of matter is from a "physical" or "hard" view and at first our purpose of matter is from this view. Matter in different forms is the building substance and constituent of the structure of hard systems. Matter with the amount of its mass and volume, determines the amount of occupied space by the structure of the related system. A structure can be made from one or various types of matter. Matter also is used for relations among the elements in systems.

But the word "matter" in addition to its physical and hard concept, also has a "logical" and "soft" concept. For example, in English language "matter" is used also for "subject" (as in "what is the matter?" or the "teaching material") that are "logical" or "soft" when compared with the hard concept mentioned above.

As we saw, the words "object", "structure", "behavior" and "order and rule or discipline", besides using for physical or hard things also are used for logical or soft things. But like the matter case, at first, our attention is on the physical or hard things. By the aid of information, we can deal with logical or soft things like the physical or hard things. If "energy" can explain the "matter" (that is more basic than to energy), then "information" can explain the "matter" and "energy" (that are more basic than to information) and in consequence the logical and soft things.

The word "energy" has its root from the word "energeia", with the meaning of "force for change". "Energy" is the resource needed for performing behavior and making change in the form of doing "work". Doing work is possible by the aid of energy and the events that occur for a structure and its resulted behavior. In other words, energy is a resource for doing behavior by a structure. If we consider matter as compressed energy in a structure with constancy, then the energy can be considered as outspread matter in a behavior with variation.

Energy can be existed in different forms. One of the main forms of energy in our world is the light and the heat resulted from it. If we eliminate the light and heat form the world that we live in, although the structures may remain as before, but most of behaviors will be eliminated from the world. Normally, form our view in the case of frigidity for example in the absolute zero and no behavior condition in physics, also information loses its importance.

The importance of the light of sun and also the fuel in the form of oil and consequently the vast needs of modern societies to electrical energy in the contemporary world is an apparent and strong reason for the role of energy and especially the electrical energy in our world. The importance of light and heat can be understood from the reality that the plants react against the light and high level living beings like human beings or animals have a mechanism for regulating and keeping the temperature of their bodies.

The word "information" comes from "in+formare" with the meaning of "to form". "Information" is a resource or agent for identifying and explaining the objects and events. Objects have their own space based properties and events have their own time based methods in the form of information. In the case of an unknown object or event, there is not any information.

By the aid of information, we can find responses for the questions like:
- "What?" including "what thing?" (noun\ object\ structure) or "what action?" (verb\ event\ behavior)
- "How?" including "how a thing?" (adjective\ property related to an object\ order in a structure) or "how an action?" (adverb\ method related to an event\ rule in a behavior)
- "Why?", "where?", "when?" and some others that some how are related to the objects or events

In a general concept, in the case of an object, "constancy" leads to some kind of "certainty" or "confidence" and "variation" leads to some kind of "uncertainty" or "doubt". In other words, in the case of an object, "constancy" leads to the validity of information or certainty and confidence. In contrast "variation" leads to the loss of the validity of information or uncertainty and doubt in the structure and behavior of the related object.

Information as the descriptor of structures and director of behaviors, describes the orders of structures resulted from matter and directs the rules of behaviors resulted from energy. This leads to "protecting the structural order and controlling the behavioral rule" or having "discipline" that governs an object in a general sense. Information is also a resource that objects use it in order to achieve their "goals".

In the case of an object that its discipline is important in order to achieve some desired goals, usually its structure and behavior can be evaluated. To do this, its structure and behavior can be compared with a balanced structure and equivalent behavior as its normal structural order and behavioral rule needed to achieve the goals while having stability. If any incompatibility or deviation from the goals found, then corrections are made to pull the structure and behavior to the desired order and rule. In human beings and animals, this is done internally and in continuous or analog form. But in most ordinary human made devices, this is done externally and in discrete or digital form.

It must be added that in the case of transcendental or superior systems like human being, the disciplinary aspects or protecting the structural order and controlling the behavioral rule that are based on information, are very important. This type of systems do not need to continuous input of matter or energy to provide their needs, but need the continuous input and process of current information from the environment by their sensors.

Evidences show that in information needed systems achieving to "stability" and maintaining it, at least needs two general sets of information including "master data = fixed basic information" and "transaction data = variable current information". This is why these two sets of information are as the base of the most information based systems. For example in a banking system there are fixed information about the account of customers and variable information about interactions between the accounts and customers. In library systems there are fixed information about the books and library members and variable information about the borrowing and returning of the books by members. In inventory systems there are fixed information about the goods and variable information about their input and output.

To protect structure and its order in systems needs to know the structural attributes or properties of the system in the form of master data or fixed basic information. At the first glance, information is the facts that identify the structure and its order in the object or system. To control the behavior and its rule in systems needs to know the behavioral attributes or properties of the system in the form of transaction data or variable current information. At the second glance, information is the facts that identify the behavior and its rule in the object or system.

Fixed basic information or master data is in two types. In superior systems, some part of master data usually is in an instinctive or institutionalized form, like what is stored in the genes. Other part of master data, according to the type of system, may be provided by training or acquisition and gradually or in one step. Also in man made devices, some part of master data is in the form of initial settings that like what stored in genes, are in the form of "factory settings". Other part of master data in man made devices is in the form of "initial settings" that are adjusted by the user of the device according to the conditions in the environment.

Transaction data is needed continuously and especially when the system behaves, must be gathered from the environment continuously. Advanced man made devices are able to gather, process and use the transaction data from the environment by the sensors that the most common of them might be a thermostat.

In other words, the requisite for stability in a system in its normal situation, is that in any supposed state in its structure or behavior, the information related to the current state must be compared with some predefined or the previous state information. This action that is some kind of "control" may be "internal control" or acted by an agent out of system as "external control".

As we see, the three main and basic resources are related to each other and in the chain of need to these resources, no resource is ignorable. But it can be said that according to the type of system, at least some level of necessity to these resources are vital. In other words, in one side, the need for matter is more vital than to energy and energy more to information and in the other side, information is more vital than to energy and energy more to matter.

In the list of resources, each one is a pre-requisite to the other. Indeed, energy needs matter and information needs matter and energy. In other words, there must be a structure in order to have a behavior and there must be a structure and behavior in order to go and find a discipline and maintain it. Study of the human kind evolution on the earth also shows this priority. The human kind has evolved from tool making with matter to think making with information.

In summary, matter is the structural substance of the objects. Energy is the ability of doing work or activity by the objects in the form of a chain of events as behavior. Information is the resource needed for protecting order in the structures and controlling rule in the behaviors while the events occur. Matter is touchable or objective. Energy if is not touchable but like light or heat is sensible or objective - subjective. Information is understandable or subjective. As we see, in the chain of the three main and basic resources of matter, energy and information, sensibility of energy is the mid point of touch ability of matter and understandability of information.

As we saw, an object with a certain structure of matter and events occurring for it due to energy, in response to the occurred events, usually behaves in some way. The structure of the object has its own describing constant data and the behavior of the object as response to the events, creates its own directing variable data. Since indeed the study of an object is studying its structure, behavior and discipline, then this constant and variable data can be used in studying the "organization" of the structure and "management" of the behavior.

It can be said that "organization" is the "structure and its order and behavior and its rule in order to achieve certain goals". Also, "management" can be understood as "protecting the balanced structural order and controlling the equivalent behavioral rule for achieving the goals". Indeed, management tries to attain some type of "certainty" by protecting the "balance" in "constancy" and "equilibrium" in "variation" while protecting the structure and controlling the behavior for achieving the goals in a best possible condition. Management also tries to promote the current level of "balance", "equilibrium" and "certainty" to a higher level, if possible.

Historical Backgrounds of System Concept

Based on what is told in the above and the subject of this book, the methods of identifying the objects, related events and their orders and rules or disciplines can be summarized as follows:

- Analyzing the whole into the parts and then studying the parts in order to identify the whole and in contrast, synthesizing the parts into the whole in order to identify the properties that can not be found in the individual parts of the whole. By studying different objects and their similarities, the "structural properties" of the objects can be extracted.

- Any event is related with a process like "action leading to reaction – reaction due to action", "cause leading to the effect – effect due to the cause" or in a general form, "inputs leading to the outputs – outputs due to the inputs". Events usually are related with objects and are identified by the behavior of their related objects. By studying the objects and their related events, the "behavioral patterns" of the objects can be extracted.

- Structures have orders and behaviors have rules. Orders in the structure and rules in the behavior of the objects due to their related events can be explained in the form of an "object protection and event control mechanism".

In short it can be said that in the case of objects and related events, usually their "structure", "behavior" and "discipline", including their "structural properties", "behavioral patterns" and "structure protection and behavior control mechanisms" are studied. As we saw and will see in the next chapters, today this is done under the system and systems thinking concepts.

In human thinking history, "reductionism" or "paying attention to the constituent parts of the whole" from a structural view and "holism" or "paying attention to the whole consisted of parts" from a behavioral view always has been the subject of discussions. In approaching to complex things, "reductionists" emphasized on the "separated parts and object as a set of parts from structural view" and in contrast, "holists" emphasized on the "integrated parts and object as a whole from behavioral view". Due to the relation between the concepts of "part" and "whole", these two methods were related with and affected each other.

"Reductionism" and "holism" have very deep roots in the history of human thinking. From reductionism view, "part" was important, because without the parts in the form of a structure, a "whole" could not be existed. But from holism view, the whole was greater than the sum of its parts, so "whole" was important and without the whole, no functionality, application or behavior was possible, because functionality is possible only by the whole and not by the separated parts.

In the past, reductionism was preferred to holism, because the normal way of identifying the things, was analyzing it into its constituent parts. The old and famous proverb of "divide and conquer", while considers the strength of the "integrated whole" (due to its emergent property by considering the relations among the parts) against the weakness of "isolated parts" (due to the ignoring the relations among the parts), indicates the preference of reductionism against holism.

In contrast, the tendency of human beings to be together in the form of societies from the most primitive to the most civilized indicates the preference of "whole" against "parts" in a natural way. In a whole, indeed when the parts relate together to form the whole, then the "wholeness" signifies the parts and the relations among them.

For example, consider several separated points or lines. In this case, there is not a single shape other than what they are. Now if the points are connected or the lines are intersected to each other in some way, then there might be a single shape. It can be said that also in a system, it is the relations among the parts that unify the parts as a whole.

The trace of these thinking methods about the part and whole also can be followed in natural languages. In language, "analysis" is to "emphasize on constituent words in a sentence" or "determining the differences and structure" and "synthesis" is to "emphasize on the whole of a sentence formed of words" or "determining the similarities and behavior".

If in reductionism parts are important and in holism whole is important, based on the standards of today in the case of human societies, while the society as a whole is important, also the individuality of the persons as parts is also important. This is the same view that system and systems thinking represent it. In other words, in system and systems thinking in contrast to the two old views of reductionism and holism, while the whole is important due to the wholeness of the constituent parts, also the parts are important.

In reductionism in that the goal was to identify the constituent parts of the objects, normally the structure or outward of the objects was important. In consequence, reductionism prescribed some kind of "seeing the outward" or the "exterior". In reductionism also some kind of "top-down analysis" was in action. The priority of parts in reductionism, led to the majority of the "structure" and "structuralism".

In holism, with the importance of whole instead of parts, normally the inward or the functionality or behavior of the objects was at the center of attentions. In consequence, holism prescribed some kind of "seeing the inward" or the "interior". In holism some kind of "bottom-up synthesis" was in action. The priority of whole in holism, led to the majority of the "behavior" and "behaviorism".

"Structuralism" mainly was used in culture, language, literature, anthropology and psychology. Also "behaviorism" especially was used in psychology. "Behaviorism" in psychology believes that all the functionalities of systems are aspects of their behavior. So, the psychopathic or psychological disorders can be decreased or removed by the changes in the behavior.

Besides these two basic methods, also there have been efforts to find the "orders" in the objects and the "rules" in their related events. In this case, the idea was that the order and rule is as a base for every thing. It must be noted that discussions about the "utopia" as a society with an ultimate form of order and rule in the works of ancient philosophers like Plato, displays this tendency among the ancient thinkers. Believing on the order and rule in the whole universe also can be seen in the works of later scientists like Leibniz or Newton.

In the beginning, thinking about the orders and rules in the objects and events, especially protecting and controlling it in the case of whole universe, went back to metaphysics. But later, due to the advent of new sciences and in the form of disciplines that sciences presented in different fields, the role of metaphysics gradually was decreased. Indeed, science is "information based" and different sciences discuss disciplines in the different fields of objects and related events that any science wants to explain it in its related field in some way.

Efforts made to discover the disciplines in the objects, led to find out how to "organize the structure" and the effects of exterior on interior and how to "manage the behavior" and the effects of interior on exterior. Efforts finally led to "organicism". Based on organicism, the whole due to the discipline has a kind of top-down domination or governance on the structure and behavior and it can be seen in the living beings in its highest form. The belief that "the whole is greater than of its parts" points the domination or governance of the whole on the parts.

Organicism, also believes that in living beings, not the structure and behavior of the constituent organs, but the structure and behavior of the living organism as a concrete and goal seeking whole is an organic process. Organicism emphasizes on the organization in the structure and management in the behavior in order to achieve the goals.

Also according to the philosophers' view, any objective appearance or subjective imagination of a thing has a "form" or "exterior" and a "content" or "interior". "Form" presents "constancy" and is as the separator of the things from each other. "Content" presents "variation" and is as the connector of the things to each other. "Form" implies some "static" and apparent exterior and "content" implies some "dynamic" and hidden interior.

The form as the exterior appears directly or sensibly and the content as the interior based on the form, appears indirectly or understandably. Since the structure has some aspects of form and behavior has some aspects of content, structure displays the form and behavior displays the content of the object. In consequence, noun, object or structure has some aspects of exterior or form and verb, event or behavior has some aspects of interior or content.

In the process of cognition, first the external forms and then the internal contents are considered. Cognition of the forms based on the exteriors and its constancy and separation of the objects from each other, leads to the classification of the objects. Cognition of contents based on the interiors and its variation and the connection of the objects to each other, leads to the discovery of relations especially the "cause and effect relation" as a common and universal tool in explaining the reality.

In the past, the objects and events were considered separately from each other and the orders of the objects and rules of the events were undiscovered. Also the causes were looked for in the out of the physical world or in metaphysic. But after discovering the cause and effect relation, man found that any event while is reflected in the structure or exterior, also is reflected in the behavior or interior of the related object. Also there is order and rule or discipline in this structure and behavior.

From ancient times, the cause and effect relation has been the most important relation in the case of objects and their related events. According to the cause and effect relation, any effect comes from an effecter. After an effect, normally the effecter of this effect is traced. Man has found the complex cause and effect relations gradually from the simple ones. Suppose the cause and effect relation in a simple form like transition of an object from existing state to a new state. Then some internal or external events or a combination of these two can be considered as the "cause" of the transition of the object from the first to the second state as the "effect".

From man's view, in similar conditions, similar causes led to similar effects, could be resulted from similar events. When an event repeated in different spaces and times, efforts were made to discover its cause and effect relation and discipline governing the case. Also, when an event repeated regularly in the case of an object, indeed it had some aspects of the discipline of the object. Discovering orders and rules in the objects and their related events had some aspects of science.

To approach to an object and explain it, its external and internal properties must be studied and the relations between its form and content must be found. According to the case, man in some cases deals with the forms and in some other cases deals with the contents of the objects. In other words, thinking is based on forms or contents. Systems approach is based on both form and content with the priority on the content.

In the past, always there were difficulties in using reductionism and holism methods separately. Some of the difficulties were from the one-dimensionality of the methods, for example, considering only the parts and not the whole and vice versa or separating the objects from the events and their orders and rules. Also only the object, event, order or rule was considered and the environment that objects, events, orders or rules were in was not considered.

Introducing the word and concept of system, unified the two old methods as a new concrete method of thinking in dealing with the complex things. The new method implied "seeing the things from the view of external form and constituent parts as structure and order, internal content and organized whole as behavior and rule and finally the discipline and mechanism for protecting the structural order and controlling the behavioral rule with the priority on whole and interior against the parts and exterior and behavior against structure".

The properties of a "system" as an "organized whole" are as "emergent properties" that the individual parts can not have it separately. "Emergent properties" result from relations among the parts and a degree of order in the structure and rule in the behavior or interactions with the environment. "Emergent properties" usually appear in a level of wholeness. Properties of systems may be deeply different from the properties of their elements. From this view, properties are either "micro" (related to the elements) or "macro" (related to the whole system). Properties of the elements or the system may vary or not vary in different spaces or times.

In consequence, a researcher in dealing with an object can act in different ways sufficient for the purpose of the research. The first method is to study the object as a "static" thing and only from the structural and its order view. In the second method while the object is as a "static" thing in some cases, in other cases is as a "dynamic" thing with behavior and its rule.

The static view has its roots from seeing the objects as "single and separated" things. But, dynamic view has its roots from seeing the objects as "organized and related" things. Finally, the researcher besides the structure and its order and behavior and its rule, also can study the ways of organizing and managing the structure and structural order and behavior and behavioral rule. Researcher may consider the above cases separately or in combination with each other.

Man has started the cognition of the objects from their constant structures and established geometry (that implies the shapes of the objects or the properties of the shapes), to study the spatial shapes of the objects. Man then has tried to quantify these properties by arithmetic (that implies constants). Man then established algebra (that implies variables) to describe the variable behavior of the objects as their qualities. This led to transition from "quantity" to "quality" or "what" to "how" in the case of objects and their events. The amount of the subjects related to variation compared to the amount of the subjects related to constancy in mathematics, indicates the importance of behavior against structure.

The advent of differential and integral calculus can be understood as finding a tool to describe the complex behavior of the objects related with the complex events.

Indeed, differential implies breaking a whole into parts and integral implies connecting the parts to make the whole. The importance of this in mathematics is due to its relation with the part and whole. Also the advent and development of different sciences can be understood as the development of disciplines in the objects and their related events in the various fields of human knowledge.

Also from the view of the subjects of this book, man while living on the earth, have tried many ages. These ages can be classified in three main eras related with the three main and basic resources of matter, energy and information. In other words, from a general view, the struggle of human kind on the planet has continued by using different dominant resources in different times as follows:
- Matter age: Pre-industrial era, matter – driven age
- Energy age: Industrial era, energy – driven age
- Information age: Post-industrial or contemporary era, information – driven age

Domination of each of these main and basic resources in the human's activities in each era, led to high improvements in human's life on the planet. Based on this, it can be said that the main dominant resource in "pre-industrial era" was "matter", in "industrial era" was "energy" and in "post-industrial era" is "information". As it is seen in this classification, domination of energy as the necessary resource for behavior was very important and led to industrial revolution.

In the case of matter, first different forms of matter like stone, wood and metal were used to make the objects and in its continuation to the contemporary era, extraction and making or synthesizing new materials have attracted attentions. Also in the case of energy, first different forms of energy like the pressure of water in a height for driving the water mills and the motion of air in the form of wind for driving the wind mills or running the boats in seas and in its continuation to the contemporary era, oil and electricity were used as the runner of the behaviors of structures.

In the case of information, first the origination of language and writing, use of clay tablets, inscriptions in the slope of mountains or use of animals' skin and the invention of paper and printing device for recording and transferring information has occurred. Then and in its continuation to the contemporary era, we are faced with communication and information technology including invention of telegraph, telephone, radio, television, communication satellite, computer and internet.

Anthropologists believe that early human beings first started making simple tools by the use of simple materials like making primary stony or wooden tools and then by generating energy from matter like making fire by wood and melting the metals, made complex tools. Human being then by accumulating the information about matter, energy and living environment, finally developed to be an idea - maker being or thinking human as homo - sapiens. It must be said that the archeologists also classify the human's ancient ages on the earth, according to the stone, metal and other similar matters. End of stone age is the time that human being started to use the thermal energy to extract metals and enter to the metal age.

Study of the struggles in different eras shows the historical roles of the three main and basic resources of matter, energy and information in the life of humans on the planet. Based on this, in the pre-industrial era, appearance of energy first led to extracting of primary metals and then led to the use of steam power that led to the industrial revolution. This development moved man from matter age to energy age. In following this development, the use of electrical energy led to huge and amazing developments in the human's life on the earth. By inventing communication devices and computer, this time man moved from energy age to information age. The outcome of energy age is different types of engines or motors that are made of different materials and use different kinds of energy. Also in the information age, inventing of the computer is like the inventing of the steam engine in the beginning of energy age. Invention of computer is the beginning of the way that must be continued in the information age.

Industrial revolution and development of the technologies based on the energy, replaced the "physical engines" or motors instead of human or livestock force in doing the work. But the information revolution has replaced or will replace the human logical or mental power with "logical engines" or motors. Industrial revolution products included matter and energy driven "hardware" with "physical or sensible aspects". But the information revolution products, in addition to "hardware", also have information driven "software" or "logical or understandable aspects".

Today, without the "physical engines" or "hard motors" like the electrical engines embedded as the "heart" in many devices around us that circulate the energy in the whole device like the blood in the body, our daily life is impossible. Also in the future (or even now), thinking about a world without "logical engines" or "soft motors" like the processor and software embedded in the devices as their "brain" that circulate the information in the whole device like the neural signals in the body, is (or will be) impossible.

In some cases, logical engines even can replace the physical engines. This replacement now has occurred in the devices like the home audio or video players. The new devices of this kind, instead of physical engine, now come with a logical engine. Logical devices also can be used as the director for guiding the physical devices in a complex way and free humans from involving in the complex situations of directing or guiding the physical devices. Here, electricity is as the blood transferring energy and electronic pulses are like the neural signals transferring information in human body.

Also it is said that the developed societies, by the use of information technology facilities, are moving from hardware based industrial societies to software based servicing societies. Information technology makes it possible, in an easy and the best ever possible way. Organized accumulation of the scientific and technological knowledge in the form of software in developed countries and transferring the production of hardware from these countries to the developing countries, are signs of this movement. In the past this transition was difficult and highly depended on the expertise of the involving workers. Application of the information technology in industry and service in recent decades has led to huge improvements for humanity, never seen in the past in history.

In summary, human kind from the time of appearance on the earth, by interacting with the "objects" in the world around, has identified the "structure" of the objects. Then by promoting the cognition process about the "matter" and using it as the resource for making the structures, has founded the first "basic sciences". Man then by discovering the relations among the "objects" and "events" has identified the "behavior" of the objects. Then by promoting the cognition process about the "energy" and using it as the resource for running the behaviors, has founded the first "basic technologies". Man also by studying the orders in the structures and rules in the behaviors, have found the "discipline" of the objects in the form of "newer sciences and technologies" and finally have identified the applications of "information" as is today. In other words, man by starting from objects and their related events, now have come to more ordering and ruling them.

In the case of "matter", establishing the networks for production and distribution of matter in the form of roads in the land, shipping lines in the sea and airlines in the air are important. Like this, in the case of "energy", networks for production and distribution of energy in the form of electrical energy networks are important. Today, in the case of "information", networks for production and distribution of information is the top subject in the societies.

In the pre-industrial era, the centrality of matter led to "structure" with constancy and silence. In the industrial era, centrality of energy and mainly the oil and electrical energy led to "behavior" with variation and motion. Today and in the post-industrial era, the centrality of information, like what is done by the information technology or in the internet, has led or will lead to more "discipline" in the human societies. But the full transition from industrial era to post-industrial era or industry to service with the governance of software in structures and behaviors has not occurred yet.

At the end of this section, it may be interesting to say that even in the case of war it was the "energy" due to the explosion of gunpowder that made a revolution in the war and improved it from what was in the past. Just like this, the use of "information" and the technology behind it, is now changing or will change the war. This will change the nature of war in the modern world.

System, General Systems Theory and Systems Thinking
In English dictionaries, various roots for the word "system" can be found. For example, internet versions of "American Heritage Dictionary of the English Language" and "Oxford Dictionaries", point to "systema" from Latin that means "to be together" and "sustema" from Greek that means "to put together" and also "synhistanai". In "synhistanai", the "syn" like the "co" in the beginning of the words, means "together" and "histanai" means "setting up", so "synhistanai" means "setting up together". Based on the many interpretations in these dictionaries about system, system is set of related things with special arrangement acting in special method.

System is a popular word and is among the 1000 most common words in English language. The word system is used in all languages and in all fields of science and technology or in general in all fields of human knowledge. The word also is used in every day conversations or writings of a wide range of people in different levels.

During last century, many new words came into existence and were used widely. Some of the words, like radio, telephone or television were the name of the device that widely used by many people. But others used only in their specific fields. Among these new words, the word system that implies a concept and is used for a wide range of things has a special situation. High usage of the word system is due to the reality that in the world around, instead of "isolated and separated parts", we are faced with "organized and related wholes".

The concept of system that introduced in the middle years of 20th century was due to the efforts made to overcome the difficulties of the ancient methods in identifying the things. By introducing the system concept, a universal method was found to identify any thing in general as a whole consisted of the related parts with special arrangement and acting in special method (from organizational view) and the necessity of protecting and controlling this arrangement and method (from managerial view).

Other reason for the generality of the system word or concept might be that it was used consciously or unconsciously to point to a natural approach to the things that was in use even before that. Indeed, introducing system concept theorized this natural approach and gave a name or title to it. By this, system and its related concepts became as a "theory for every thing".

System concept also introduced the "system environment" concept and replaced the traditional view of "independent whole" by the new view of "system and its environment". From this view, system is a set of related parts as a whole that interacts with its parts and environment in an organized and purposeful method. Also "environment" is the world around that system lives in.

The word and concept of system is also some how related with the word and concept of "set" as a collection of things in mathematics. According to "set theory" in mathematics, a "set" is a collection of things each as a "member" or "element" of the set. In mathematics, "set" is also very common and in wide use. Like considering all the things as a system from our prospective, "set" is also used to define or describe many concepts in mathematics.

In the case of a set, it is not necessary that all the elements forming it really or physically gather in one place. Of course it must be clear that a certain thing belongs or does not belong to a set. In other words, when we can not determine that certain things belong or not belong to a set, then such a set might not be existed. This is also true about the system and its elements as well.

Acceptance and development of these two concepts (system and set) in their related fields is reasonable. Here we have replaced the "set" with "whole" or "system" and the "member" with "part" or "element". When we talk about a specified set, we can say that certain things are or are not in this set. This is also true in the case of a system and its elements. As we will see, this is more in the case of hard systems and less in the case of soft systems. Like sets in mathematics, systems also can have elements of different types.

The advent of set theory also implies that in the real world, instead of "isolated single" things, we are faced with "related multiple" things. This contains the same concept we mentioned about the system earlier. Also the concept of "universal set" or "set of sets" that is known as "U set" in mathematics, is due to the reality that the things we deal with, not only are as wholes but also are as parts of bigger wholes and in its ultimate form the whole universe. The whole universe contains all the others and is the environment of all the systems.

In modern mathematics, the generality of "relation" and especially the "function and variable" is due to the importance of constancy in the objects and variation in the events in the real world. From mathematical view, system can be defined by "functions" or equations and especially differential equations. Each function can display one aspect of the system. A hierarchy of systems also can be defined as "nested sets" or "function of functions". Of course the concept of system like the concept of set tends more to the concept of "quality" than to the concept of "quantity".

The concept of system is explained in "General Systems Theory = GST" and in a framework that is called "systems methodology". Like other theories, GST includes general principals independent of a certain object, event or order and rule. GST is a set of principals that are used to explain a system and its situations in the past, present and future. Also the "systems methodology" is a set of methods and tools that is used in systems research.

As the background for GST, we can say that the scientific developments in 20th century changed the Newtonian mechanical view about the world and objects and their related events and orders and rules. Scientific development revealed that "matter" and "energy" are two forms of the same thing. During the 20th century, also the "information" as the third resource was added to the two other known resources that were "matter" and "energy".

The appearance of information and communication theory and the developments begun by the end of 19th and the beginning of 20th century that led to the invention of telegraph, telephone and radio, changed the face of the world in the 20th century. As a result of developments, a new era started in that information had an effective role and the subjects like "organization and management" and "organizing the structure and managing the behavior" became the fundamental and important subjects.

Also it was found that in living beings "feed-back of information" is related with the protection and control and this process supplies the required information for the living being. Understanding the "relation between living being and living environment" led to the advent of new concepts. 20th century also was the time of appearance of huge technological complexes for production, distribution and consumption of matter, energy and information. Hence, it was required that instead of seeing the objects, events and their orders and rules separately, to see them as wholes consisted of many related parts.

The advent of GST and systems thinking based on it that was useable in many fields, played an effective role in the cognition process. GST is a unique interdisciplinary theory. Also the systems thinking is a universal thinking methodology for studying the complex objects, related events and the orders in the objects and rules in the events. GST is used in all fields of science, technology and the whole knowledge field of human beings.

GST changed the traditional mechanical method that only dealt with hard objects and promoted it to a new method for dealing with both hard and soft objects and their related events. In other words, GST can be used to study the hard, soft and hard – soft objects. In a general view, GST has philosophical, scientific and technological aspects.

In the past and in classical sciences, the mechanistic and analytic method in studying the objects and their related events only considered the unidirectional flow of causes and effects in the events. But, GST while considering the bidirectional flow of causes and effects, with a holistic and syntactic view not only considers the objects and their related events but also the orders and rules in them. GST also emphasizes on the organized and purposeful whole and its management, environment and its effects on the objects and events.

According to GST, biological societies are systems with interacting elements that have mutual relations with each other and the capability of self regulation in structure and behavior. The biological societies also have the capability of adaptation with their environment and maintaining some kind of stability in the form of balance in structure, equilibrium in behavior and certainty in discipline.

GST with centralizing on the wholes consisted of parts, relations among the parts and the role and importance of environment in the whole and the fact that each whole is a part of a bigger whole, affected all the scientific and technological fields emerged from the developments in 20th century. Parallel to this, continuation of information revolution emerged from the wide spreading use of computers and internet at the end of 20th century, totally changed the face of the world.

As we saw, considering an object as a system in a certain boundary, needs to look for finding a certain type of structure, behavior and orders and rules as discipline for it. According to GST, always only those elements must be considered as the elements of a system that relating them with each other contains some kind of wholeness with certain goals or purposes to achieve. What makes this wholeness and goals or purposes, is the view or the goals or purposes of the researcher studies it.

When the goals or purposes of the researcher change, then the elements of the system and environment also change and instead of the previous elements, other new elements must be considered. Considering a boundary for systems, causes to have a criterion for belonging or not belonging of certain elements to the system or its environment and identify clearly what is ongoing as behavior.

"Systems thinking" is a combination of "achieving from the parts to the whole" and "achieving from the whole to the parts" and is used in recognizing the objects, their related events and orders and rules as their disciplines with emphasizing on wholeness. When we talk about behavior, then also the goal is argued. Goal as a destination to achieve is meaningful when there is a behavior. Also when there is a behavior and a goal to achieve, then order and rule in the structure and behavior that implies discipline is also argued.

Systems thinking, in addition to emphasizing on the "parts", "relations" among the parts and the "whole" resulted from the related parts as "structure", emphasizes also on the "behavior" and "discipline" in the structure and behavior. This is done by defining the "cause and effect relations" or in the universal form of "process" that includes "input – function – output" and the "environment" that the structure behaves in it. "Environment" in its ultimate form is the whole universe or a subset of it that includes the space that objects exist in it and the time that events occur in it.

The pre systems thinking era method as "analyzing the whole into the parts" that emphasizes only on "structure", can be called as "systemic approach" or "order based approach". Also "synthesizing the parts into the whole" that emphasizes on "behavior", can be called "systematic approach" or "rule based approach". The combination of these two approaches also can be called as "systems approach" or as today is used, as "systems thinking".

With this definition, "systems thinking" as a new approach, implies the two old methods simultaneously with priority on whole against parts. This approach besides the "structure" and "behavior", also considers the "discipline" or "order" of the structure and "rule" of the behavior in the whole. This aspect of approach as "organizational – managerial approach" implies the "organization and management" in the case of objects, their related events and orders and rules in them with the aid of "information".

Human Being as a Complete and Superior System
Human being has always been used as a model to describe different concepts. Indeed, human being is a complete model of the thing that is called "system" today. In human being as a living being, the silent aspect of the body as structure is made of different kinds of matter. Also the moving aspect of the body related to a set of events as behavior uses different kinds of energy. Balance in the silence aspect of structure, equilibrium in the motion aspect of behavior and certainty in protecting the structural order and controlling the behavioral rule in the whole body of human being as a superior living organism, needs different kinds of information.

This is the reason for why the human being has certain organs like digestion for "capturing, keeping and processing the food" to produce "matter" for body "structure" or "energy" for body "behavior" and other certain organs like brain and senses for "capturing, keeping and processing data" to produce "information" for body "discipline". Indeed, some part of daily food is used for restoration of the body, like generation or restoration of bones or muscles or totally the structure. Other part is used to produce and keep the energy needed in our daily activities or totally the behavior.

Some part of energy stored in body is used for the daily activities as behavior. Other part of energy is used to gather, store and process data to produce information from the living environment. This is done by the organs like eye, ear and brain. The outcome of all these, is the "balance" in structure, "equilibrium" in behavior and "certainty" in protecting structural order (as we like to be) and controlling behavioral rule (as we like to do) or totally the "stability" of the body.

Thinking about human beings also reveals some other facts in this case. The food in the form of matter or energy as "input" to our body first is stored in the related organ (stomach) and then goes to "process" in the other related organ (intestine). The input food stored in is as a "temporary storage" and normally goes out of its storage to be processed. But the energy stored in body, is a "permanent storage" and its consumption may last longer time. Both of these, some how go out of body as "output" in the form of work or garbage.

The same process for the food (matter and energy), is also valid for information. In other words, human being in addition to matter and energy also needs information about the world around. Some part of the input information like the input matter and energy by processing in the brain is used as short term memory for current activities like in walking that we bypass the barriers in our walking route. Other part of the input information as useful and necessary information is stored in long term memory in order to be used when we need it.

In today living, the importance of information for human being when compared with the past or other resources, not only is not less, but also is more. In modern societies today, the needs for primary resources like matter and energy are provided in some way. The importance of mass media and the high tendencies for knowing or getting news is a sign of this situation in our contemporary world. In other words, the human life, from all dimensions is a flow of "inputs", "storage of inputs", "processing inputs to produce outputs", "storage of outputs" and "outputs".

A chain of sequential and related actions with a certain purpose to achieve and in the form of the major steps as "getting inputs", "saving inputs + functioning on or processing inputs to produce desired outputs + saving outputs" and finally "giving outputs", is called a "process" or "Input – Functioning or Processing – Output = IFO or IPO" in summary. As we see, the human life on earth is a total process consisted of other partial processes.

It may be interesting to say that the same process is done in one of the most advanced and complex systems man ever made and that is computers and the computer based information systems. At the beginning side, computers have certain devices such as keyboard, mouse, microphone, camera, scanner and other devices (similar to eye for reading, ear for listening and others in humans) as "input devices" for getting in the data and instructions. At the central part, computers also have "main memory" or RAM (Random Access Memory) for short term storage of inputs (data and instructions) + "processor" or CPU (Central Processing Unit) for processing the input data by running the input instructions to produce output information + "peripheral storage" or HDD (Hard Disk Drive) for long term storage of input data and instructions or output information (in total similar to brain in humans). At the end side, computers have certain devices such as display unit or monitor, speaker and printer (similar to language for speaking, hand for writing and others in humans) as "output devices" for put out the information.

Thinking about the daily activities of humans, also reveals different combinations of some basic actions. According to this, the functioning or processing as the backbone of any function or process is a combination of three basic actions called "sequence", "selection" and "repetition". We can identify these basic actions in all the behaviors of humans or human made devices.

"Sequence" is the primitive and normal method of doing actions in that all the actions needed to achieve a goal, are done one after another. In other words, any action that can be done without any question is a "sequence". For example "cooking food", "eating food" and "washing dishes" are a "sequence" of actions that most of us do it every day.

"Selection" is the action of making decision to continue the actions in this or that way when there are two or more ways to continue. In other words, "selection" is choosing one of the "two" or "more" "possible ways" to continue the operations. For example, suppose that you have put a side a certain amount of money for buying some thing. When you go to buy it, if its price is near the assigned money, you may decide to buy it. But if the price is greater than the assigned money, you might not buy it. What helps you to buy or not to buy it is comparing your money with the price and then deciding "to buy" or "not to buy".

"Repetition" is an action in that certain steps are done again and again until a step wise goal to be achieved. In other words, "repetition" at its least form is a combination of "sequence" and "selection" that must be done again and again until a goal is achieved. With this description, we may think that the "repetition" is not a basic action. But since the repetition indeed means "doing again and again" and this is the building block of many behaviors, it can be considered as a basic action.

For example, suppose that you must go out of home at 11 o'clock, because some one will be there waiting for you. Now if you are ready to go out, you will wait and checking the clock until the time you must be there. The action that you do is a "repetition" consisted of a "sequence" ("waiting" and "checking the time for 11 o'clock") and a "selection" (to go or not to go). In consequence, you may still stay at home or to go out of home.

As a total and complete example of these basic actions, suppose that you want to buy few things. In this case, you decide to buy the things one after another since you have enough money in hand. In this example, buying the things (as "sequence"), deciding to continue or stop buying (as "selection") is done again and again (as "repetition") until all the things are bought.

2

System and Related Concepts

Introduction

As we saw in the first chapter, "system" is used for a wide range of objects, related events and orders and rules. "System" is also used to describe the structure and structural order, behavior and behavioral rule and protecting the structural order and controlling the behavioral rule as discipline in the structure and behavior. Properties of a system emerge from the relations among the elements of it as a whole. Difference between a system and a non-system object is that in a system object excluding a part from the whole, usually breaks its structural wholeness and destroys its behavior. But in a non-system object, there are not relations among the parts and a wholeness or behavior not considered for it.

The first step in recognizing an "object" is to know "what object is?", "what action does?" and "how is and how does?". To know "what object is?" needs to know the "structure" of it. To identify "what action does?", needs to identify the "behavior" of it. To understand "how is and how does?", needs to understand "what structure it has and what behavior it does". This implies understanding of the "structural order and behavioral rule" and "mechanism for protecting the structural order and controlling the behavioral rule" or the "discipline" governing it.

For dealing with the objects and their related events, man normally started from simple aspects and then added complex aspects to it. To describe the concept of system, this evolutionary approach is also used in this chapter in dealing with the objects and their related events as a system. Most of the concepts related to system, are explained in this chapter.

The wide range of the cases that the word "system" can be used for, is due to that it can be used for the objects, events, objects and their related events, objects and orders in the objects, events and rules in the events, or structures, behaviors and disciplines, single or in combination with each other. Using this approach in dealing with the objects and events in the world around, helps the researcher to start form thinking about the structure of the object and then about the behavior based on this structure and finally the orders in the structure and rules in the behavior as discipline. This method, as a normal approach, starts from simplicity and gradually moves to complexity.

Different Levels of System Concept

A general and universal definition of system is that: "system is any object consisted of related elements as a whole with a certain structure and because of its wholeness as an entity, certain goal, purpose, function or application can be considered for it." This general definition for the system is only from the structural view that implies "constancy" or "stasis", "matter" or "structure with stasis" aspects that implies "space".

In such an object as a system, the "energy" or "behavior with variation" aspect in "time" and "information" or "discipline as order in the structure and rule in the behavior" in "space - time" is ignored. This definition implies a "static view" of the object in the "space". So, it can be called as a "static system".

According to this definition, "static system" has constant structural properties and no events occur for it. Also because of the "structural constancy", the "structural order" of it remains the same as it was. In other words, "static system" is a "constant structure" or "single state" system and has the same structure or remains in the same state as before.

An object in its simplest form can be viewed as a "static system". Also the reason that a researcher views an object as a "static system" is that its structural properties do not change or no events occur for it. In other words, the object does not have any behavior. In the absence of behavior, the object protects its primary structural order and remains in the same state as it was in the past.

In a "static system" with no variation or behavior, any change in it, like changing the place of a piece of stone in a road by a passer's kick, is from an external source and as a response to an external event. This change is not in the nature of the stone and the time has no role in it. Because changing the position of the stone in the road does not have any meaning for the passer in the road.

The objects that are called "structure" satisfy this definition of system. Based on this, from a researcher's view, structures like Eiffel Tower or Egyptian Pyramids consisted of many elements as their parts related with each other in a certain order and totally form a "whole", can be viewed as a "static system".

The goal of making the Eiffel Tower as a structure was the symbolic aspect of it in a fare in Paris. The case also was discussed at the time of building the tower and criticized for the sake of no useful application for it. In the case of Egyptian Pyramids as a structure also there have been discussions about the applicability of the Pyramids as tombs for the Egyptian Pharaohs or astronomical purposes.

In any case, the reason that a researcher considers the objects like these as a "system" or in exact as a "static system", is their "wholeness" with a certain "structure" consisted of many elements related with each other or placed besides each other in a certain order to form the whole object. A set of elements stocked but not related with each other (like a full set of an automobile parts in a stock), without any wholeness, goal, purpose or certain functionality or applicability, does not make a system. Of course, from another researcher's view, even each of these elements may be considered as separate systems.

Using "a set of elements" in the definition of system implies that systems have more than one element (at least two or more elements) and the elements are related with each other in some way, directly or indirectly. In other words, systems have at least two elements and one relation between the elements. This means that there is not any element in a system with no relations with other elements in it. Using the word "system" for an object usually implies some kind of "getting together and having relations with each other". An object in a least form is viewed as a system when it is consisted of several related elements as a whole.

For example, a piece of stone in a road, from a passer's view that uses it for breaking walnut, may not be viewed as a system. But the same piece of stone from a chemist's or geologist's view interested in to study the molecules or atoms of the stone or its environment, may be viewed as a system.

We can say that, using the word "system" for an object indeed reflects its researcher's view. Evidences show that in the world that we live in, there is no monolithic, non-decomposable or "without a structure" thing. In general, any object even the atoms or atomic particles can be viewed as system and studied based on the system concepts.

Considering an object as a system or non-system thing from a researcher's view, depends on the presence or absence of certain properties in the studying object. The most important cases are decomposability against non-decomposability, relations against no-relations among the elements and the presence or absence of wholeness emerged from the presence or absence of relations among the elements. This reminds the old philosophical and familiar concept of "atom".

With this interpretation, approximately any object that we see or imagine in the world around us, are as systems or parts of a bigger whole or elements of a bigger object that with other parts or elements, form a system. Based on this, "systemism" is the belief that "any thing is a system or an element of a system".

From a researcher's view, when an object is not decomposable or has no parts and in consequence there is no relation among the parts and finally no wholeness, is not helpful to be considered as a system. For distinction between a system and non–system thing, we can use the words "object" or "entity" against the word "system".

As a result, considering an object as a system depends on its researcher's view and from this view, is the researcher's "system of interest". "System of interest" is an object with certain goals and purposes or functions and applications that based on some reasons a researcher likes to study it as a system. In this case, system concepts may be used to study it.

It must be added that "system" mostly is as a general or universal frame that can used in a gradually growing concept as explained in this chapter. However, some theorists for distinction between what is and what is not in the domain of a certain study, in addition to the word "system", use the word "holon" too.

Holon is a neutral form of the Greek word "holos" that means "whole". Holon is used for "part" or "element" and "whole" or "system" or any thing that is a system or non-system thing. The neutrality of holos means that it can be used for "part" or "element" and "whole" or "system" at the same time. In other words, any thing that is or is not a system from a view can be called "holon".

A combination of all the cases involved in a "view" to "a specific thing in a specific point of space and moment of time", make a specific cognitive "perspective" about that thing. Like in painting that the perspective changes when the "viewer", "space" or "time" is changed, the cognitive perspective about a thing also changes when the researcher, space or time is changed.

In the cognition process of a thing, in different points of space and moments of time, there may be persons with different perspectives involved in. Here also like in painting, the number of perspectives in different points and moments, can lead to a better view and finally to better cognition of the thing.

But in the real world, we are not faced only with the objects as "static system" and the above mentioned definition about the system is not sufficient. To have a better and more complete concept of the system and in consequence be able to deal with a wider range of objects, in addition to the "matter view" or "structural view with constancy" in the above mentioned definition, we must also consider the "energy view" or "behavioral view with variation" due to the related events with the objects.

Hence, because of the "energy" and "variation and behavior", in addition to "space", also the "time" becomes important. With this more complete concept of the system, in addition to the "structure" which is important from "constancy" or "static" view, also the "behavior" which is important from "variation" or "dynamic" view, is considered. In consequence, in the studying object, in addition to "matter" or "structure" and "space", also the "energy" or "behavior" and "time" becomes important.

This definition about the system that implies "dynamic view" besides the "static view" is more compatible with the objects and events in the real world around us. Such a system can be called as a "dynamic system". A "static system" (like a seed of a plant with the rootlet, plumule and food reserve) in a suitable environment (like the wet soil with sufficient temperature), can promote to be as a "dynamic system" (like a plant).

For example, objects with a type of "mechanism" for behaving imply this definition of the system. In consequence, things like a clock (one of the first most effective man made machines) or an ordinary water supply set with some type of mechanism for doing their work can be considered as "dynamic system".

It is interesting to say that the "clock" as one of the first man made systems with some type of behavior, was made to measure the "time" as the "yeast" or the necessary thing for any "change". Clock needs energy for behaving (moving the arms for seconds, minutes and hours). In the past, energy was supplied by twisting a spring inside the clock, sufficient for a limited length of time. In the primary clocks, the cogged wheels as parts were related orderly with each other. The full set of the clock could use the energy as "input" to produce the moving arms as "output". The "goal" of the clock as a "whole" was to measure the time.

Also an ordinary water supply mechanism has specific parts for the "input", "reserving + filtering" and "output" of the water. The parts are related with each other with a specific order as the structure of water supply set. The structure, by "input, reserve + filter, output" the water under a specific rule as the behavior, makes the water supply mechanism as a "whole" for supplying the water as a "goal".

The reason that a researcher can consider an object as a "dynamic system" is that from the researcher's view or in the domain of the research, the object of interest behaves in time in some way. The object may display a smooth or different behavior in time and may be in different states in different times. When behaving, also the structural properties of the object may remain constant or change in a certain way. In the first case we are faced with a "constant structure" and in the second with a "variable structure" during behaving.

As we see, unlike a "static system", in a "dynamic system" the structural properties in space may remain constant or vary in time. If during the behavior, the structural properties of a dynamic system remain constant and the system behaves in the same way in time, then we are faced with a "steady state system". For example, a moving thing with a constant speed in a straight path in space out of gravitational forces and no changes in structure and behavior can be viewed as a steady state system.

Also a system like a magnetic compass (with the energy resource from the magnetic poles of the earth) that reacts against the changes (like moving it around) and maintains its former state is a "state maintaining system" or an "adaptive system". But indeed, a "dynamic system" is a system with "state transition" and in any moment of time may be in a different state. In a dynamic system, in addition to "matter" for "structure" in "space", also "energy" for "behavior" in "time" is needed.

Behavior of dynamic systems less or more obey the specific models like "cause and effect" or IFO or IPO models discussed in the previous chapter. Indeed, dynamic systems accept certain inputs and by doing their natural or defined function or process on it, produce certain outputs. In other words, dynamic systems "do behavior due to inputs to produce outputs".

When system behaves, inputs are passed to the elements that can handle it in order to do "operations" or "initiate or terminate" the behavior of the elements. In this process, outputs of some elements may act as inputs to the other elements. The final outputs of the system are the result of this chain of mid-inputs and outputs. The consequent behavior of different elements of the system that makes the "system total behavior" may be very different from the behavior of the separate elements.

Again, for a more complete definition about the system in order to approach with a wider range of objects, we must add the "information" or "order in the structure and rule in the behavior" as "discipline" and the "mechanism for protecting the order in the structure and controlling the rule in the behavior", to the above mentioned definitions. Note that the above mentioned definitions implied the "matter view" or "structural view with constancy" and the "energy view" or "behavioral view with variation" due to the related events in the objects. With this final definition, now we are faced with objects as system in that the structure and structural order in space, behavior and behavioral rule in time and how to protect the structure and control the behavior in space – time or discipline is also important.

We can say that "discipline" in relation with "structure" and "behavior" implies some aspects of "protection and control" in "space and time". From this view, the meaning of "discipline" is "protecting the structural order in space" and "controlling the behavioral rule in time". In other words, "organizing" the structure and "managing" the behavior in space and time requires "information".

"Organizing the structure" and "managing the behavior", especially when is done by system itself, implies some kind of "being alive" and can be seen in the living beings or "organisms". This definition of system also implies an "organizational and managerial view" about the objects in the world around us. Hence the system of interest can be called as "organic system".

With this complete and final definition of system, in addition to former definitions, we can consider all the living beings including humans, animals, plants and most of the new man made technological devices, as systems. This definition includes a wider range of objects in the world around us and is more compatible with the real world.

As we see, the "organic system" is a "dynamic" and "multi state" system. Organic system for "structure" in "space" requires "matter", for "behavior" in "time" requires "energy" and for "organizing" the structure and "managing" the behavior in "space – time" requires "information". "Organic system" indeed is a "dynamic system" in a "variable environment". It means that, with the "changes in the environment in a normal form", organic system can protect its "balance in structure, equilibrium in behavior and certainty in discipline". In other words, it can "protect the structural order and control the behavioral rule" or totally maintain its "stability" and remain as a "stable" system.

The survival of an organic system in a normal condition, needs getting "inputs" from the environment in the form of three main and basic resources including "matter", "energy" and "information". The function of the system is "saving, processing and consuming" the inputs by the "organs" or "elements" to produce outputs. Finally the "outputs" in a useable or waste form again in the form of three main and basic resources are given to the environment. In this total process, also "organizing the structure and managing the behavior" must be done continuously. In other words, despite the various changes or transformations during the behavior of an organic system, the survival of the system requires some kind of "stability" by protecting the "balance" or structural order and controlling the "equilibrium" or behavioral rule and having some kind of "certainty" or "confidence" in these cases.

As we said, "stability" is a desired or favorable state and systems resist against the changes in order to keep it. The requisite for keeping the "stable state" in an organic system is structural "balance" from matter view, behavioral "equilibrium" from energy view and disciplinal "certainty" from information view. The sum or total outcome of all these is "stability" in systems.

With the changes in the environment, the "organic system" by getting the necessary "inputs" from the environment can change its structure or behavior so that the supposed wholeness in the structure or behavior remains as before. In consequence, the system with a degree of "stability" can "stand still" or remain to be as it was. Such a system is called a "homeostatic system" that includes all the living beings and "organisms".

In other words, a "homeostatic system" is a system that is "self structure protector" and "self behavior controller" in a specified range and in consequence a "stable and firm" system. Now we can call this kind of systems as "self organizing and self managing systems". In this kind of systems, the required changes in the structure or behavior of the system and protecting the structural order or controlling the behavioral rule are done by the system itself or its internal and environmental agents.

"Organization" in general, is a system with certain goals that is properly structured and properly behaves to achieve the goals. The "proper structure and behavior" means having a specified order or arrangement in the elements besides each other and a specified rule or method in the functions that the system does in order to achieve its goals. "Organization" can modify its structure or change its behavior if the internal or external conditions change. By finding a new proper structural order and behavioral rule in the new condition, can adapt itself with the new condition and continue to exist. Also "organizing" is the process of ordering the elements and ruling the actions so that the whole set of elements is properly structured and properly behave in order to achieve the specified goals.

In an organization, at least one element has the responsibility of monitoring or control in order to achieve the goals of the organization. "Organization" is different from "organism". Constituent elements of an organization while having a whole structure and behavior in order to achieve the goals also are meaningful and useable separately. But in organism only the whole has meaning and the parts are not independent of the whole and can not be used separately or in other cases. In living beings, the organs (like the hand or foot of humans), have such a complex and interwoven structure and behavior so that when separated from the whole, are destroyed and become useless. For this reason, the individual organs can not be used separately or it is impossible or very hard to use it in other similar systems.

In systems important from structural or structural and behavioral views, usually the disciplinal view might not be the case of the attention. In this type of systems, protecting the structural order and controlling the behavioral rule may be done by external agent like a person responsible to it. This task may be added to the systems by modifying the structures of the systems like what we see in the new electronic doors when compared with the old ordinary doors.

Living beings as superior systems existed on the planet from ancient times. Living beings use all of the possible combinations of matter, energy and information as inputs and by functioning on inputs produce the desired outputs, necessary for living on the planet. What we know about the systems today is resulted from the struggles man made to continue and promote living on the planet and also by thinking about the objects, their related events and orders and rules.

First understandings of the systems, started from understanding the structure of the objects. Then besides the structures, the behavior of the objects attracted attentions. Finally, understanding the mechanism of protecting the structure or structural order and controlling the behavior or behavioral rule followed. Man, in understanding the behavior, functions or processes, started from matter interactions and then studied the energy and finally the information interactions in the objects.

In simplest forms, systems can be considered only from the view of "structure", "behavior" or "discipline". In complex forms a combination of these three like, "structure and behavior", "structure and discipline", "behavior and discipline" or in complete form "structure, behavior and discipline" can be considered. Indeed, in different studies, different aspects of the system may be the subject of interest, although it may be necessary to consider other aspects too. All the objective or subjective entities with some kind of "structure", "behavior", "discipline" or a combination of these three, can be studied by using system concepts.

In different positions or discussions, elements of systems, may be pointed as "part", "object", "entity", "component", "member", "thing" or "element" in general. Remember that considering a thing as an element of a system, depends on the researcher's view and the goal of the research.

In this book, we have used "whole" against "part" and "system" against "element". The constituent elements of systems can be "hard\ physical", "soft\ logical" or a combination of these as "hard\ physical - soft\ logical". Hence we can say that physical elements as "hard" or "concrete" elements include "hardware", logical elements as "soft" or "abstract" elements include "software" and physical - logical elements as "hard - soft" or "concrete – abstract" elements include "hardware – software".

Like this, the relations among the elements and the inputs and outputs of systems, can be classified in the above mentioned broad categories. Hard relations are physical or hardware, soft relations are logical or software and hard – soft relations are physical – logical or hardware – software dependencies among the elements of the system.

Thinking about the set of elements that makes a system, leads to think about the relations and interactions among the elements and the ways that a system as a whole relates with another systems as another wholes. "Systems thinking" implies that any part of a whole considered as a system, is understandable only in relation with other parts of the whole as system.

As we saw, basic properties of systems emerge from the relations among the elements and the wholeness of system. Changes in the elements or relations among the elements in a system may damage this wholeness and in consequence damage the system. Then the whole set is not the same as it was in the past and we can not consider it as the same system.

The simplicity or complexity of systems depends on the certain aspects that we may consider in them. The aspects include structure, behavior and discipline. In general, considering systems only from structural view is simpler than from both structural and behavioral views. Also considering systems from structural, behavioral and disciplinal views simultaneously is more complex than the systems with lesser aspects. Diversity of elements, number of relations among the elements (from structural view), diversity of inputs, functions or processes and outputs (from behavioral view) and diversity of mechanisms for protecting the structure and structural order and controlling the behavior and behavioral rule (from disciplinal view) are other factors of simplicity or complexity in systems.

In a normal condition, when the diversity of elements and the number of relations among the elements from structural view, diversity of inputs, functions or processes and outputs from behavioral view are less and the mechanisms for protecting and controlling from disciplinal view are understandable, then the system is also simple. If not so, then the system might be a complex system. In a big or complex system, elements, relations among the elements, inputs, functions or processes, outputs, orders, rules and even the system goals can be classified in different classes.

Evidences show that in the list of the three main and basic resources and system types, "time" has relative importance. In other words, in the "matter", "energy" and "information" list of resources, with matter there is less dependency and with information, more dependency on time. This is also true in the list of "static", "dynamic" and "organic" systems. Indeed, with static systems there is less dependency and with organic systems, more dependency on time. Importance of time increases from matter or static systems to energy or dynamic systems and finally to information or organic systems and decreases in reverse.

In other words, in "static systems" or systems important only from structural view (like the peace of stone in our example), only the material structure of the system is the subject of interest. For studying the behavior, then we must consider "energy" and "time" in it. Like this, when we want to discuss about the order in the structure and\ or rule in the behavior, in addition to matter and energy, we must consider information in it. In consequence, in such a system, for "protecting structural order" and "controlling behavioral rule", in addition to "matter" in "space" and "energy" in "time", also "information" in "space and time" is required.

Starting from "silence" implying "matter" in "space", first we are faced with "static systems" and some kind of "infiniteness" in time. Then we have "motion" with the aid of "energy" in "time" and "dynamic systems" and some kind of "finiteness" in time. At the end we have "silence and motion" with the aid of "information" in "space and time" in the "organic systems" and some kind of "urgency" in time. For example in the case of human beings as organic systems, if "matter" is needed in longer periods of time, "energy" is needed in shorter periods and "information" is needed in any moment of time.

Because of the importance of time in dynamic systems when compared with static systems, static systems can be classified as "time invariant" and dynamic systems as "time variant" systems. Also, some kind of "life" and "length of life" or "age" can be considered for systems. Like before, this life in one end is "infinite" and the in the other end "instant".

For making an object, the maker first makes the structure of the object. Then, after adjusting the structure, thinks about adjusting its behavior. Finally the maker thinks about mechanisms for protecting structural order and controlling behavioral rule. In other words, the natural way for making objects, is to start from structure, then test the behavior and finally add the structural protection and behavioral control mechanism to it.

This is the way that for example a carpenter makes an ordinary door. Carpenter first makes the frame and leaf of the door. Then connects the leaf to frame by hinge to complete and adjust the structure of the door with its specified order. Then tests the opening and shutting as the behavior of the door and if necessary, makes corrections in the structure until the door behaves correctly. Finally adds the knob and lock to it to complete the disciplinal aspects of the door.

Today, we can see how the man made systems operated "manually" in the past, turn to operate "automatically", like what we see in the old ordinary and new electrical doors. Also some kind of protection and control mechanism in man made systems that in the past was based on "repair" and activated in system fails now is based on "maintenance" and is active to keep the systems out of fail.

"Automation" and "maintenance" are among the most important subjects in the industry and new systems. Indeed, "automation" and "maintenance" are related with "protecting the structural order and controlling the behavioral rule" in systems. This leads to "continuous protection of structural order and control of behavioral rule with no or least human involving".

System Environment Concept
"System environment" is the set of elements not belonging to the system, but affect the structure and structural order and behavior and behavioral rule of the system and also the system affects it. In addition, system usually gets "inputs" from and gives "outputs" to its environment. Usually, the environment of a system physically is besides or around the system, but may be far from it.

Environment of a system as a "context" is a bed that the structure of the system is "depended" to it, the behavior of the system "flows" in it and the system "reacts" with it. The result is certainty in protecting the order and maintaining the balance in the structure and controlling the rule and maintaining the equilibrium in the behavior. Internal elements of systems have more effective role in systems than the environmental elements. Because systems like living beings, have mechanisms to adapt the internal elements with the changes in the environment.

Environment of a system may provide "facilities" or create "restrictions" for survival of the system. If a system requires new facilities in the environment, some times the required facilities are already available and some times it must be added to the environment. Also some times for the further survival of the system in new conditions, system must be transferred to a new environment.

The type and speed of changes in the environment of a system is another factor in the adaptability of system with the environment and survival or destruction of system in the new conditions. Structural balance, behavioral equilibrium and disciplinal certainty or "stability in the system", needs "stability in the environment".

Systems are separated from their environment by a boarder that is called "system boundary". Indeed, "system boundary" determines the boarder of the system from their researcher's view. Also the "wholeness" and "goal" of the system find their meanings inside the system boundary. From different views, the boundary of a system may be different.

Assuming boundary for systems, leads to the identification of the system from its environment and in consequence the whole as system. This also leads to considering a "domain of definition" for the system. Viewing system in its environment, separated from the environment by its boarder, can help the researcher to have the "big picture" of the system (system with its environment) against the "small picture" of it (system without its environment) in mind.

To determine what is "in" and what is "out" of the system always is not simple. This depends to the researcher and the boundary assumed for the system. Researcher may assign some elements to the inside or outside of the system. When the researcher's view changes, then the internal and external elements of system may be substituted. In other words, with extending the boundary, the former external elements may become the new internal elements of the system. Unlike this, by shrinking the boundary, the former internal elements may become the new external elements of the system.

For example, in the case of "hard systems" like an automobile, from an engineer's view, the boundary of the system is clear and certainly does not include the driver. But from a traffic expert's view and in the case of "soft systems" like the behavior of the drivers in traffic, the view may differ and also the driver may be an element of the system.

Possibility or impossibility of studying the elements as internal or external elements restricts system and non-system elements in the researcher's mind. If study of the element is possible, then there may be interests to consider it as system element. Otherwise, there may not be any interest to consider it.

Systems Hierarchies

Determining boundary of systems, leads to "system domain" and hierarchies of "sub-domains" concept. Domain of the system is really the realm or territory of the system that includes the structure and behavior and disciplinal protection and control takes place in it. System domain may be argued for the structural, behavioral or disciplinal effects of the system and from this view may be limited.

Systems have their own domains. Big domains may be divided into several sub-domains with smaller realms or territories. In this case, the upper domains can use the facilities of the lower domains. Each sub-domain may represent a "sub-system" under the main domain as "main system".

As we know, "system structure" is the "set of system elements" and "relations" among them. "System behavior" is the "function of system structure" on "inputs" to produce "outputs". "System discipline" includes the "orders in the system structure and protecting it" and "rules in the system behavior and controlling it" with the aid of a "protection and control mechanism". Like this, the individual elements of systems may have their own structures, behaviors and disciplines.

When each of the individual elements of a system satisfies the concept of system mentioned above, then each individual element can be considered as "sub-system" under the big system as "main system". In any case, any system in a specified domain can be identified by analyzing the structure, synthesizing the behavior and describing the structural order and behavioral rule and the related protection and control mechanism.

Any system can include several sub-systems. Each sub-system may be in its sub-domain under the main domain in relation with other sub-systems in the set. In this case, sub-systems can be imagined as "super elements" of the main system. Also "meta element" as an element that defines other elements and "micro\ macro element" as an element that is "small\ big" compared to the ordinary scales are among the words that may be used in this case.

If a system as sub-system is an element of a bigger or main system, then the bigger or main system can be considered as the "environment" of this sub-system. A system consisted of several sub-systems is called a "super-system". Also "meta system" as a system that defines other systems and "micro\ macro system" as a system that is "small\ big" compared to the ordinary scales, are among the words that may be used in this case.

Super element is a macro element in a system consisted of other micro elements and different from the others but with no tendency to consider it as a "sub-system". The word "super element" is used when a whole structure is broken in to partial structures and then there may discussions about considering them as sub-systems or elements (now as super element) in the whole structure. Also super system is a macro system consisted of related super elements as micro or sub-systems, like a set consisted of sub-sets. The word "super system" is used when there is a set of systems and the researcher wants to consider them as a macro system.

If sub-systems contain each other one after another, then we have a "hierarchy of systems". In a hierarchy of systems, systems in different levels interact with each other. In this hierarchy, lower level systems are sub-systems of higher lever systems. Also the higher level systems can use the facilities or outputs of the lower level systems. The "goal" of a sub-system usually is highly depended on the goal of the main system that includes it or is dictated by the main system.

Systems with common environment may interact with each other through a common section. This common section like the dashboard in an automobile is as a point for getting together of sub-systems and is called "interface". "Interface" is a common place for different systems and connects systems with each other. As we see, a system may be a sub-system of a bigger or main system and at the same time may be as a big or main system that includes other sub-systems. The concept of "hierarchy of systems" implies the "layered structure" in systems.

In a layered structure system with a hierarchy of systems, the place of each system is in a specific level that may be higher or lower when compared with another specified level. In studying this kind of systems, the researcher always is faced with some kind of "inclusion" or "supposing a system inside the other" and "exclusion" or "supposing a system outside the other".

Also we may face with a "linear chain of sequential systems". These systems get the initial inputs in one end and give the final outputs in the other end. Processing initial inputs to produce final outputs is done by the mid-systems to produce the mid-outputs. For example a group of industries or factories that get raw materials in one end, then produce the semi-manufactured and finally the full-manufactured products in the other end, are examples of this chain of systems.

The complete chain of the above mentioned systems may be viewed as a main or "total system". In this case, this main or total system gets the inputs as the total input, functions as the total function and gives the outputs as the total output. In this chain of systems, systems interact with each other through matter, energy and information interchange. Indeed some of the outputs of initial or mid-systems are entered as inputs to the mid or terminal systems.

Systems Types and their Inputs and Outputs
As we saw, "matter", "energy" and "information" are the three main and basic resources in systems. Based on this, inputs to systems can be in the form of "raw matter, energy and information". Like this, outputs can be in the form of "transformed or processed matter, energy and information" in a simple or very complex form.

According to the list of three main and basic resources and the importance of each resource in different systems, the following fundamental classification about the systems can be introduced:

1. "Matter driven or static systems": Systems with matter as the main resource, "structure" or "construct" important only from structural views.
2. "Matter and energy driven or dynamic systems": Systems with matter and energy as the main resources, "mechanism" or "behaving structures" important from structural and behavioral or only from behavioral views.
3. "Matter, energy and information driven or organic systems": Systems with matter, energy and information as the main resources, "organism" or "organization" important from structural, behavioral or disciplinal or a combination of these (structural and behavioral, structural and disciplinal, behavioral and disciplinal or structural and behavioral and disciplinal) views.

Systems with interactions in the form of "input, function and output" with their environment, are as "open system". Otherwise, especially with no inputs from the environment, are as "closed system". Usually, "dynamic or organic systems" are "open" and "static systems" are "closed" systems.

In an open system, the environment of the system plays an important role in the behavior and discipline of the system. But, in a closed system, theoretically, environment has no decisive role. Any supposed interaction in closed systems is internal and among the elements of the system. Static or structural systems are typical example of closed systems. Dynamic systems need external sources of energy for behaving. Also organic systems in addition to energy also need information. Dynamic and organic systems are typical examples of open systems.

In closed systems, knowing the properties of system at a moment of time helps the researcher to forecast the future of the system. But in open systems, because of the many environmental factors, forecasting the future of the system is very complicated and hard or impossible. In closed systems, ignoring some parts of the system might not destroy the system. But in open systems, ignoring some parts of the system might destroy the wholeness and damage the behavior of the system.

Like the elements of systems discussed above, also the systems can be "hard\ physical" or "soft\ logical" or a combination of these as "physical – logical" or "hard - soft". Like elements, we can say that physical systems as "hard system" or "concrete system" include "hardware", logical systems as "soft system" or "abstract system" include "software" and physical - logical systems as "concrete – abstract system" or "hard – soft system" include "hardware – software".

In the list of resources and systems, from matter to information or from static to organic systems, hardness decreases and softness increases. Also in the case of complexity in systems, the degree of complexity increases from closed to open systems and from hard to hard – soft and soft systems.

Entropy and its Concept in Systems

"Entropy" as an "inverse scale to display the presence or absence of energy for doing work", first derived from the second law of thermodynamics. The laws of thermodynamics are related with gases, heat and motions in machines as the cornerstone of industrial revolution. As we will see in this section, the concept of entropy in systems is very comprehensive.

For example, heating water causes motion in the molecules of water. When the water is warm, the motion is also high and when the water is cold, the motion is also low. In consequence, when there is enough energy to heat the water, then it is possible to create motion by consuming the energy.

Consider some water warmer than the environmental temperature. If there is no energy to keep it in the same temperature or to heat it up, then its temperature will fall down gradually and finally will be the same in the environment. In this case, with no energy as input to heat the water, unlike the decrease in the temperature, the entropy of the water will increase. Finally when the water temperature falls down to the temperature of the environment, then the entropy of water will be in its highest value.

If there is enough energy to keep the water in the same temperature or increase it, at least for a length of time in the future, we can keep the water in the same temperature or increase it. When the energy ends, again we come back to the same state as before. Now, with energy as input to heat the water, the entropy of water like the temperature of it will remain the same or will decrease to its lowest value.

In the above mentioned example, entropy in reverse is equivalent to the motion or silence of water molecules. In other words, when there is energy, the molecules of water have more motion, so entropy remains constant or decreases. Unlike this, when there is no energy, then the molecules of water have less or no motion and finally a degree of silence. As a result, entropy increases to its highest value.

As we see, the basic concept of entropy is also related with the central subject in the systems or motion, variation and behavior with the aid of energy. Later, the concept of entropy has been used as a criterion for "certainty" against "uncertainty" and "order" against "disorder" in the information theory. In closed systems isolated from the environment, entropy remains constant or increases. But in the open systems, by interactions with the environment, entropy may remain constant or under some conditions may decrease.

A comprehensive definition of entropy can include the three main and basic resources of matter, energy and information and also the structure, behavior and discipline. In other words, it can be considered as a factor of presence or absence of structure (matter), behavior (energy) and discipline (information). In this case, it can be as a variable between 0 (maximum presence of each case) and 1 (maximum absence of each case). Entropy can be viewed as a criterion or scale for "weakness or strength ness" from structural view, "silence or motion\ stasis or dynamism" from behavioral view and "order and rule\ regularity or disorder and misrule\ irregularity" from disciplinal view.

From this view, we can evaluate entropy in all systems. In a normal condition, entropy is like "time" and as a total scale for "lifetime" or "age" tends to increase by time. Based on this, if a system "is left to be as it was" or a closed system with no inputs in the form of matter, energy and information, tends towards "weakness or losing relations" from structural view, "silence or stasis" from behavioral view and "disorder and misrule" from disciplinal view and the entropy of it increases. But if a system "is not left to be as it was" or an open system, by interacting with the environment can slow down the normal increase of entropy.

Indeed, there is no system to remain in the same state or condition for ever. Evidences show that with the unidirectional increase of time, in a natural trend, all the systems objectively or subjectively tend towards some kind of structural destruction, behavioral silence and disciplinal disorder and misrule in general. Finally the system may change its state or reach to an end as it was.

It must be noted that while in the open systems entropy can be decreased, but this decrease is accompanied with the increase of entropy in its container system or environment. In other words, entropy of the whole universe as the container system or environment of all of the other systems steadily increases and this total increase may be in the form of decrease or increase in other systems it contains. For example if human activities lead to decrease of entropy in some systems, but this decrease also leads to increase of entropy in the living environment as the container of other systems.

More entropy indicates more clutter or disorder in the structure and more silence or misrule in the behavior. When we say that a system is in its "stable" state, it means that the system is in a most probable sustainable state from structural, behavioral and disciplinal views. In this case, the entropy of the system remains constant.

In order to remain in the stable state, systems need to receive the necessary inputs from the environment. Inputs to the systems, lead to resistance against the increase of entropy. If a system does not have any interaction with its environment and can not receive the necessary inputs to remain in its stable state, then the entropy also tends to increase. As we noted above, the result of this condition is losing the desired structure, behavior and discipline in the system.

For better understanding of the concept of entropy, imagine a building that is maintained by periodically repairing with suitable substances as input and kept against the natural destructions. In the case of this building as a system, the increase of entropy will be very slow. Unlike this, if the building "is left to be as it was", the entropy will increase and by destruction of the building, will reach to its highest value. In the case of this building as a system, highest value of entropy indicates the highest degree of losing the structure.

Structural balance, behavioral equilibrium and disciplinal certainty or in total, "stability" of system, are "maintained" by input, processing inputs to produce outputs and output. Matter, energy or information as inputs, are consumed, transformed or processed to produce the desired outputs. Wile there are sufficient inputs for system needs, also the structural balance, behavioral equilibrium and disciplinal certainty or stability of system can be maintained. As a result, also entropy can remain in a constant value. In other words, entropy is related with "balance in structure", "equilibrium in behavior" and "certainty in discipline" in systems. We can say that the total entropy in systems is the average of the entropies discussed above about the structure, behavior and discipline.

When the necessary inputs to the system decrease, system may keep its stability for a time in the future and the entropy may remain in the same level. But with interruption in inputs, system will fail and the entropy will increase. Entropy is also as a scale for "youthfulness" and "oldness" or "life" and "death" in systems. In a general view, survival of systems is a result of structural balance, behavioral equilibrium and disciplinal certainty or stability of the system. This requires keeping the entropy in its lowest level.

According to this view, we can use the concept of entropy to describe our ordinary "good feel" or "bad feel" in our body as a system. When we have "good feel", then the entropy is low. Unlike this, when we have "bad feel", then the entropy is high, if compared with the entropy in good feel. The highest value of entropy in this case is in "death" time or when our body as the structure of system decays and turns into soil.

In the above discussion, if we come one step down from "energy" to "matter", then entropy is also related with the potential of transforming matter into energy. If we go one step up from "energy" to "information", then entropy is also related with the potential of transforming energy into information. As transforming or consuming of matter to produce energy may need its necessary inputs, then transforming or consuming of energy to produce information also may need its necessary inputs.

Organic systems and human kind societies as open systems become more organized and gain more order and rule by the time. This means that entropy remains constant or decreases by time. Open systems have their own mechanisms to fight against the normal increase of entropy in order to gain more stability and promote their current situation.

If we consider the maximum value of entropy equal to death, then living beings as organic systems usually have their own natural mechanisms for fighting against the death or increasing of entropy. This kind of systems by getting the necessary inputs in the form of matter, energy and information from their environment can keep the level or fight against the increase of entropy in order to gain more stability.

In any case, living beings as organic systems have a limited length of life time and by the end of this time, inevitably die with the highest value of entropy. But, living beings and especially human kind societies, with the aid of information, can transfer the old orders and rules from one generation to another and promote it by time and gain more stability. In consequence, as the time goes on, if they have the necessary abilities, they can experience "growth" and finally "evolution". If they have not the necessary abilities, they normally are forced towards "destruction" and finally "extinction". In the first case, entropy remains constant or decreases and in the second case entropy increases.

Resources (especially matter and energy) in the environment are limited and the open systems use it to continue their life. If the amount of the sources decrease and the used sources not supplied again or some kind of saturation occurs in the environment because of the high number of users, then different scenarios are possible. If the used sources not supplied again or the environment saturated by the users, there may be "slow down or stop of the growth leading to evolution" or "speed up or intensify of the destruction leading to extinction". Even mutations may occur in the users leading to exit from the conditions ruling the environment. In the case of supplying the used sources and preventing the saturation, we may face with a "closed repetitive cycle of birth and death".

In other words, when sources in the environment are not enough or available, then the entropy in open systems will tend to increase. Under this condition, also the process of input and output decreases or stops and system goes towards some kind of destruction or death. But when sources are enough or available, then the system can continue its normal life. As a consequence, the entropy of the system remains as before or decreases to lower levels. Now if we generalize the above mentioned cases to all systems from simple to complex and resources from matter to information, then we can have the following conclusions.

If we consider the simplest systems as systems only with material structure with no energy and information, then the existence of matter can help the system to keep the same relations and orders in structure and keep entropy in the same level as before. But, even in such systems, as a result of continuous effects of the environment on the structure of system, entropy normally increases with time though very slowly.

For example, consider a building on our planet earth. If this building built with the best material and the strongest structural relations, despite the use of all the best against destruction, in a normal condition, it will tend to destruct and entropy increase. But if we suppose such a building on the moon, then according to the conditions ruling the moon as the environment of the building, the above mentioned state may not be the case and in consequence entropy may remain in the same level as before.

One of the ways to overcome this normal destruction and increase of entropy in the systems with structural aspects is to apply some kind of "maintenance" or continuously strengthening the material and structural relations in the system. This is exactly the work that is done continuously on the structures like Eiffel Tower in Paris. In static systems like an ordinary building, in that only the structural aspect of the system is important, entropy can remain in the same level as before or increase. In this case, entropy is a scale of keeping or losing the structure and structural orders in systems.

Now we go one step forward and consider more complex systems that are important from both structural and behavioral views and the two main and basic resources including matter (needed for structure) and energy (needed for behavior) with no attention on information. In this case, the existence of matter and energy can help the system like the moon in its orbit around the earth, to protect the structure and structural order, control the behavior and behavioral rule and continue its behavior. As a result, entropy that may vary in a range of values in different times can remain in a certain level or in a normal value in average.

In this case, continuing the behavior of the system needs to get inputs (as raw matter or energy or both), do the behavioral operations of the system on them (by consuming or processing some part of the matter or\ and energy input to the system) and to give outputs (as processed matter or\ and energy). In other words, survival of this kind of systems in time depends on a continuous stream of getting, processing and giving of matter or energy or both. The type and speed of this stream, depends on the type of the system.

For example, consider an early transforming manufactory as a system with a certain mechanism to transform a simple elementary thing to a simple secondary thing. This manufactory can keep the entropy in a constant level only in its normal condition that is to get, transform and give the matter and energy by its order based structure and rule based behavior. Otherwise, the entropy will increase. Also, a "stable state system", when can not behave, tends to its most possible state in that the entropy remains constant or increases.

In our classification of systems in this book, the most complex systems are those that use all the three main and basic resources. As the use of "energy" in systems adds behavioral aspects, use of "information" adds disciplinal or "organization" and "management" aspects to systems. In this kind of systems, not only the structural (including matter) and behavioral (including energy) aspects, but also disciplinal or the organizational and managerial aspects are important.

Continuation of the behavior of these systems also needs getting inputs (in the form of matter, energy and information), functioning on inputs (by processing some part of matter, energy and information input to the system) and giving outputs (in the form of processed matter, energy and information). As before, survival of this kind of systems in time depends on a continuous stream of getting, processing and giving matter, energy and information. Again the type and speed of this stream, depends on the type of the system.

For example, now consider a modern industrial factory as a system. Such a factory can keep entropy in lower levels only when behaves normally by getting, processing and giving the three main and basic resources. Otherwise entropy will increase. The advent of organizational and managerial dimensions in systems, as having the abilities to organize their structure and to manage their behavior, is an outcome of adding the "information" as a new resource to the list of required resources in the systems.

The concept of entropy also can be extended to soft systems. It must be said that the human societies as soft systems, like other systems in general also obey the general principals of systems. For example the current modern societies in the form of states or governments, also have the main aspects of structure (framework of state or government as a whole), behavior (executive power) and discipline (judicial power for orders and legislative power for rules). A stable society is also a society that can keep the entropy constant or in lower levels. In this case, the most stable society is one that has the lowest value of entropy with lowest level of tension and crisis from different aspects.

The history of social changes indicates that the change is the nature of human societies. While, in animal world we can not see the same changes as in human world. Of course, talking about the change in the animal world like the change in the human world may not be meaningful. The cause of this is the third resource or information and the quality of gathering and processing of data and using the resulted information in human societies. Talking about the "information age" or the "free flow of information" that nowadays we mostly hear about, are the signs of the importance of this role of information.

Summary of System Key Concepts
From the whole material discussed above, we can extract the following keyword concepts about systems in general. The concepts include "what is" or structure, "what does" or behavior and "how is and how does" or discipline including the order in the structure and rule in the behavior:

Structure
- "Elements" as the constituent parts of the system
- "Relations" among the elements of system
- "Wholeness" or the "whole" consisted of "parts"
- "Goals" of the system
- "Environment" of the system
- "Structural properties" of systems
- "Structural order" of system

Behavior
- Actions related with the "inputs" to the system, event or action as cause
- Actions related with the "function" of the system
- Actions related with "outputs" of the system, event, outcome of reaction against action as effect
- "Behavioral patterns" of systems
- "Behavioral rule" of system

Structural order and behavioral rule as discipline
- "Structure order protection mechanism" in system
- "Behavior rule control mechanism" in system

Based on this, in approaching to the objects, related events and their orders and rules, if we can identify the above mentioned key concepts and explain them, then we have been able to identify the object and describe it. As we saw, "system" is a complete concept and most of the objects in the world around us, have system type natures. Now we summarize the key concepts about systems:

- Element: Element is any individual and separate unit as a "constituent part" in a set identified as a "system". If the structure and structural order, behavior and behavioral rule and the mechanism of protecting the structure and controlling the behavior of an element inside a system are important for its researcher, then it may be considered as a sub-system inside the main system. "Element" has general types as follows:
    1. Physical: Element with sensible material nature
    2. Logical: Element with understandable logical nature
    3. Physical – Logical: Element with sensible material nature from one side and understandable logical nature from other side

- Relation: "Relation" is any kind of dependency between two or more elements inside a system. Relation in general produces the coherence of the whole system and the result of it, is: "the whole is greater than the sum of its parts" as a system. Supposing an element belonging to a system but with no relation with other elements in it is not charming or meaningful.

- Wholeness: "Wholeness" is the integrity of system from structural view, the possibility of having behavior by the use of structure and the order in structure and rule in the behavior, in order to: "the whole to be greater than the sum of its parts".

- Goal: "Goal" is the things or subjects that explain or justify the existence of structure and doing behavior by a system. "Goal" indeed explains the "purposes" that the structure exists in order to achieve to these purposes. Goals in systems are the navigators of the structure in doing the behavior in order to achieve the purposes of the systems. Orders in structures and rules in behaviors are maintained in order to behave properly towards the goals.

- Environment: "Environment" is the set of factors that affect the structure and structural order and behavior and behavioral rule and the system also affects it. Environment of system can be a bigger system that the system of interest is as a member or element of it. In this case, input – output or interactions of the system take place inside it.

- Input: "Input" is any thing that is "entered" to the system. Input is entered to the system in order to "be used" (while preserving it like work tools), "be consumed" (while reducing it like fuels for producing energy) during the system behavior or to "initiate\ terminate" functions in the form of system behavior. Inputs can be in following general types:
    1. Matter
    2. Energy
    3. Information

- Function: "Function" is the set of actions that the system does on input in the form of its behavior in order to produce output.

- Output: "Output" is any thing that is "exited" from the system. Output is as the "product of the function of system". Outputs can be in following general types:
  1. Processed matter as the product or material waste or garbage
  2. Produced or transformed energy as a potential for doing work or energy waste
  3. Processed data as the necessary or unnecessary information

The effect of changes in the inputs on the outputs of system can include the following cases:
- Increase in inputs leads to increase, decrease or no change in some outputs
- Decrease in inputs leads to decrease, increase or no change in some outputs

Based on input and output, the concept of system can be categorized in the following main categories:
- Only structure with no input and output as a closed system
- Structure, behavior and discipline with matter, energy and information input for doing behavior while protecting the structural order and controlling the behavioral rule and matter, energy and information output as an open system.

- Structural order: Any arrangement of elements in systems that must be existed in order to have wholeness and behavior for achieving the goals. This order must be protected in order to have structural balance as an aspect of "disciplinal certainty" or stability in general.

- Behavioral rule: Any method about the behavior of system that must be existed in order to have wholeness and behavior for achieving the goals. This rule must be controlled in order to have behavioral equilibrium as an aspect of "disciplinal certainty" or stability in general.

Stability in systems:
- Balance in the structure
- Equilibrium in the behavior
- Certainty in the discipline

General system types
- Physical: Ordinary automobile
- Logical: Set of principles with specified orders and rules like banking or judicial system
- Physical – Logical: Programmable equipments or hardware with software

Structure and behavior of few systems:
Ordinary factory

- Elements: Machinery, equipments, mangers, workers
- Relations: Relations among the equipments in assembly line, among workers, mangers and workers, workers and machinery, etc.
- Wholeness and goal: Factory for producing certain products
- Inputs: Raw materials
- Function: Transforming raw materials into products
- Outputs: Products

Financial accounting
- Elements: Financial documents, ledgers and accountants
- Relations: Relations among accounts, ledgers and accountants with each other
- Wholeness and goal: Financial administration and balance of accounts
- Inputs: Financial documents of incomes and costs
- Function: Keeping the account of incomes and costs, receipts and payments
- Outputs: Ledgers and balance sheets

Judiciary
- Elements: Courts, judges and acts
- Relations: Relations among acts, courts and judges
- Wholeness and goal: Justice and maintain security
- Inputs: Claims and litigations
- Function: Judgment
- Outputs: Orders and sentences

Parts, relations, inputs and outputs in systems
Parts
- Physical: Equipments, machinery and humans or animals
- Logical: Facts, instructions and acts
- Physical − Logical: Electronic digital components with hardware and software

Relations
- Physical: Couplings, bridges and links
- Logical: References among facts, instructions and acts
- Physical − Logical: Hardware - software interface

Inputs
- Physical: Raw material and energy resources
- Logical: Data, information and problems
- Physical − Logical: Raw material, energy resources and related data, data on paper or memory device

Outputs
- Physical: Products or semi-processed material
- Logical: Information and solutions
- Physical − Logical: Products or semi-processed material and related information, information on paper or memory device

Sub-systems under some systems
Main system: Human body system
- Sub-system: Sensation
- Sub-system: Breathing
- Sub-system: Blood circulation

Organs in human body
- Structural organs: Skeleton and digestion
- Behavioral organs: Muscles, hands and foots
- Disciplinal organs: Senses, eye and ear

Main system: Government in a country
- Sub-system: Executive power for behavior in the form of whole government structure
- Sub-system: Judicial power for disciplinal orders
- Sub-system: Legislative power for disciplinal rules

Organs in governments of countries
- Structural organs: Legislative power, state cabinet and ministries
- Behavioral organs: Executive power and executive organizations
- Disciplinal organs: Judicial power, parliament, police and army

Historical trend of understanding of system concept:
Studying of the history of human activities, science, technology or inventions shows that we have come to the current concept of system in the three following steps:
- System as a structure with or without matter input, like Egyptian Pyramids as a whole with the goal of making a burial place for Pharaohs or astronomical purposes.

- System as structure and behavior and with input and output in the form of matter and\ or energy. Mechanisms like a stockroom, water reservoir, hive or crock for storing substances such as water and flour. Mechanisms like furnace, fireplace or stove for generating heat. Mechanisms like clock as the first man made auto machine with a simple regular behavior in the form of the motion of clock arms. Mechanisms like water mill or wind mill or the mill with animal forces and with the first "feed-back loops" in that appropriate to the intensity of water or wind and in consequence the speed of rotation of the upper stone in the mill, grains entered to the mill.

- System as structure, behavior and discipline and with input and output in the form of matter and\ or energy and\ or information, like large scale auto machines and factories as outcomes of industrial revolution, computers, identifying organisms as "systems" and presenting GST, huge industrial complexes and modern organizations and managements with the information technology as their backbones.

3

Systems
Structure, Behavior and Discipline

Introduction

System and related concepts were discussed in the previous chapter. We saw that from a researcher's view, an object as a "whole consisted of related parts", can be considered as a system from different views. The views include "structure", "behavior", "discipline", "structure and behavior", "structure and discipline", "behavior and discipline" and "structure, behavior and discipline".

In this book we use system in its most complete concept as a "whole consisted of related parts" with certain "structural order" or properties and "behavioral rule" or methods, that has a certain "discipline" in the form of a "structure protection and behavior control mechanism" and in order to achieve its "goals" can "protect" its structure and "control" its behavior.

This concept of system attracts the attentions from the object and related events to the internal structure, behavior and discipline and also the environment of the object. With this approach we find that internal structure, behavior and discipline and also the environment and stability in the objects that we consider as systems (in its complete form), have basic roles in them.

In this chapter, we will explain these concepts in detail. We will see that how matter, energy and information are the three main and basic resources in the structure, behavior and discipline of systems. For example a normal human being as a complete system, while needs to get matter and energy in the form of food for maintaining the structural balance and behavioral equilibrium in doing the daily activities, also needs to get information from the environment for maintaining the disciplinal certainty.

Systems Structure

As we know, based on system concepts, any object as a system, at its least form, is consisted of "structure". Structure of the object has a basic role in considering it as a system. Because, using the word system about an object, at least needs to have a structure with certain wholeness.

"Structural elements" of a system as a "whole", generally include the "parts" of the "whole" or the "elements" of the "system" and "relations" among them. In other words, the constituent elements of a system are related with each other in order to fulfill the proposed "wholeness". The main aspect of wholeness in systems is "structural wholeness". According to this view, structure of system as a whole is important either only from structural view or from the view that structure supports behavior towards the proposed goals.

Structure of a system is identifiable as a whole due to the relations among its elements. In addition to partial properties, structure also has certain general properties. In other words, structure is a combination of elements besides each other with a certain arrangement as structural order with balance and are related together so that the whole set is identifiable as a single unit with certain general properties. In this case, the set as whole can be considered as a "structured" object. Structural order with balance is a state in that the system has structural stability. As a consequence, system can continue to exist as it was.

In systems, elements can be "visible" or "invisible". "Visibility" is a structural property and determines the "visibility rate" of system. If we consider "identification" as "the ability to separate the elements from each other", then "visibility" of elements in systems implies that the elements of the system are "identifiable" with their own borders by the researcher.

If the elements or the borders of the elements in a system are not identifiable, then they are "invisible". In this case also the "visibility rate" of system is definable as "low visibility rate" or "high visibility rate" according to inability or ability in identifying the elements with their borders.

Being "replaceable" is a structural property in that the existed structural elements can be exchanged with other similar elements. Indeed, "replace ability" refers to the ability of an element to be exchanged with other element. Also the "equality", "similarity" and "symmetry" are other structural properties from disciplinal or structural order view in systems. "Equality" and "similarity" also are as means for classifying objects in different classes. We will discuss these subjects in the "systems discipline" section in this chapter.

In daily natural language conversations, objects appear with their "name" in sentences. A listener or reader, when listens or reads a name in a sentence, usually imagines an object in his or her mind. This imagination is mostly about the structure or structural properties of the object. In other words, nouns in sentences help listeners or readers to have certain related objects in mind and visualize the structure of them.

According to the type of relation among the elements, systems have different general structural properties as follows:
- "Continuous structure": A kind of structure like the structure of a static system in that only the structure of system is viewed. In this case, at the first glance, the relations among the elements are not important and the structural wholeness of system is viewed "concrete and monolithic".
- "Discrete structure": A kind of structure like the structure of a dynamic system in that in addition to the structure, also the behavior of system is viewed. In this case, the relations among the elements are important and the structural wholeness of the system is viewed "divided and modular".

A researcher, by the use of a "model" can view the "continuous" structure of a system as "discrete" structure and restrict it to the structural elements important from the research view.

As we saw, various words like "part", "element", "object", "thing", "entity", "component", "member" and some other words are used to refer to the constituent parts of a whole or the structural elements of a system. These words or idioms while may have different meanings in different cases, but in the case of system all have the same meaning as "a part of whole". It must be noted that elements in systems are important due to the "roles" they have in the system and not the words used to refer to the elements.

Among the many words or idioms noted above, perhaps the "part" against the "whole" and "element" against the "system" is more convenient. Also, traditionally "part" is used against the "whole" and "element" against the "system". Here, first we will have a short explanation about the "part" and "element" and then about the "object" and other words and idioms.

"Part" is used to refer to separate constituent things of a whole inside it. "Whole" is opposite to "part" and is used to refer to a set of related parts in that the parts complete each other to form a thing as a unit. From this view, "a part from a whole" is a "piece", "segment", "division", "portion", "fraction", "fragment", "section", "constituent" or other similar cases of the whole.

"Part" is smaller than the "whole" that includes the part. From a mechanistic view, the sum of the "parts" in combination with each other, create the "whole". But from systems view, "whole is greater than the sum of its parts". Whole is created when all the constituent parts of it are related together in order to organize a set of related things. Wholeness brings "quality" from the "quantity" of related parts.

"Element" is any fundamental or basic thing that may be considered individually or in relation with a system. When an element is considered in relation with a system, then it may be viewed as a constituent part of it. The concept of element as a part of a whole some how implies no attentions on the details of it at the first glance. An element inside a system may be viewed as a sub-system inside the main system.

The word "object" as a thing that may be a system, a part of a whole or an element of a system is also used in different scientific fields. Scientists refer to "objects" in their scientific studies. Like "element", "object" is also a general idiom that is used to refer to some thing that the details of it, is not important at first. "Object" may include a wide range of things in the world around us or the whole universe. "Object" is one of the common words that is used to refer to different things in general or in the field of science and technology.

When an object is identified or defined, then some of the "attributes" or "properties" of it may be noted. From a researcher's view, the noted cases usually are some how "quantitative" (can be measured) or "qualitative" (can be evaluated). Objects can be "physical" as a "sensible" or "real\ actual" thing or "logical" as an "understandable" or "virtual\ imaginary" thing or a combination of these two.

The word "object" also widely used in information systems field. In information systems, "object" is an abstraction of similar "things" with their common "attributes" or "properties" and can be described in similar way and each considered as an "instance" of that similar things. Here concept of object implies an unspecified instance but common among the similar things as the whole set of instances. The set of the similar things is considered as a "class" of those things. Classes also can be physical, physical – logical or logical.

In information systems, classification of the objects in different classes, reveals the basic relations among them. Usually an object becomes identifiable when we can put it in a class of similar objects. In this classification, identifying of the object includes "identifying the class of the object" and "identifying the instance of the object in its class". As we see, here object is single in word but plural in meaning.

Also "property" of an object includes "a simple and one item description" or "data" about it. Objects usually have some basic and general properties like "name" and "identifier". "Name" refers to the object among the other similar objects. "Identifier" is a kind of specific property that only refers to a specified instance. Objects may have one or more identifiers, like first name, last name and unique identification number of a person. Usually, one of the several identifiers of an object, like the unique identification number that identifies the object specifically, is preferred among the others and is called the "identification key" of the object.

A property (like size, color and etc.) of an object (as an abstraction of similar things with similar properties), is an abstraction of a property among the several instances that the object abstracted from it. Properties have specific names. In the case of different objects or in different points of space or moments of time and in the case of a specific object, properties may accept different values. The values can be "quantitative" or "qualitative".

In a class of objects, if this value is "quantitative", then it is in the form of a "numeric value". In this case, usually the value has a "range" with its maximum\ minimum or includes a list of certain cases that the object can take one of the values or cases in different spaces and times. If it is "qualitative", then it is in the form of a description with a specified "concept".

"Thing" is another word or idiom that is used to refer to the parts of a whole. When we can not find other word to refer to a part of a whole, then we may use the word "thing" for it. In this case, it means that we do not know any more about the nature of it. If we knew it, then we could find other word to refer to it.

"Entity" is any "thing" that comes from a real or supposing "existence". This word is from the Latin root "ens" (existence). Our imagination about some thing as an "entity" really is the imagination of "existence" of it. We use "entity" to refer to any thing that "really exists or supposed to be existed", and as a consequence, we can deal with it. In other words, in order to deal with some thing, it must be "existed" in some form.

When we use "entity" for some thing, indeed we want to "separate" it from other things and "define" it. When there is not a name or a description about an "identifiable" or "definable" thing, then we may use "entity" for it. Here "entity" is a general concept and can be used for "whole" or "part" and "system" or "element" or the things in the "environment" of a system.

"Component" is a thing that is in relation with other thing. Here, the placement of a component in its right place and related with other things to make the whole is important. In the absence of a component we may not have the whole or the desired function. Usually, a component in a system is replaceable with a similar thing. In other words, component can be separated from the system and replaced with other similar thing and relate in the same way with the other components in the system.

"Member" is an individual and implicitly "unique" thing that is in relation with a "set" or "group". Using "member" as a part of a whole, implies that the member is besides or related with other members in the whole. A member as an organ in an organic system (like the heart or brain in human beings), may be very important in the whole, but in other systems (like members of a library) with no more importance in it. Organs like the human being organs may be very complicated and have interwoven (organic) relations with other members. This is why, when they are separated from the whole, become useless or using them in other places is impossible or very hard.

When the parts of a whole can be viewed independently in their own separate positions, then from structural view we have an "aggregation". "Aggregation" is "coming together" and implies some kind of "multiplicity". Unlike this, when the parts of a whole are not meaningful without the whole then from structural view we have a "composition". "Composition" is "merging to each other" and implies some kind of "unity".

As we saw, properties of a system as a whole are the outcome of structure, behavior and discipline of system. For example, if we gather all the necessary parts of a car in a place and even arrange the parts like a car but with no relations among the parts, we will not have a car. A car is made only when the parts are related with each other just in a way that they "must relate" so that the car to behave normally. "Relation" among the parts is one of the systems necessities.

At its least form, relation among the elements can be defined by putting them in the same class or category of elements. Same as this, we can define relations among the various classes or categories of elements. Of course, in the case of structure and especially physical and hard structures, usually the physical relations among the elements are at the center of attentions.

By relation among the elements and the cause and effect principal, any change in an element related with other element causes the change in the other element too.

Also by the cause and effect principal, with change in the second element, there may be changes in the first element. This change may continue until reaching to a new "balance" in the structure in new condition.

In order to show the importance of order in the structure and relation among the elements in a system, consider an ordinary door. An ordinary door is consisted of "frame" and "leaf" and relation as "hinge". In order to have a "door" as a system, first the frame must be fixed vertically balance in the wall. Then the leaf must be "related" with frame by the use of "hinge" so that to find its right place inside the frame. Now we can have the supposed behavior of the door in the form of shutting or opening it when necessary.

In the above example, when the hinge does not act properly, the frame is not fixed in vertical balance in the wall or the leaf does not fit in the frame, then the whole door will not behave properly. In our words, in this case "system fails". Now we add "knob" and "lock" to it as disciplinal aspects. In the door, like in an "organization", the "knob" and "lock" facilitate the "management" of the door or open\ shut and lock\ unlock it in different times. Knob and lock also act as user interface.

In this example, frame, leaf, hinge, knob and lock related with each other and wall, make the "structure" of the door. "Behavior" of the door is to open or shut it by the use of an external source of energy or the arms of the user of the door. "Discipline" of the door from structural and behavioral views, is balance of the door in the wall and opening or shutting and locking or unlocking it in different times by the user.

In an ordinary condition, it is desired that the door to be closed and when necessary, to be opened. In this case, adding things like spring or electrical door opener may facilitate the behavior of the door and managing of it with or without the aid of a human agent. We must confirm again that the behavior of an ordinary door is possible by the aid of external energy source of the arms of a human agent as its user. But in modern electrical doors, energy needed to operate the door is supplied by an electrical motor and the disciplinal aspects of the door, facilitated by the electronic sensors.

Like the system elements, for the relation among the elements in systems, different words or idioms are used. These words or idioms while are different, all imply the same meaning as "to be in relation with". Among the many words, "relation" may be the most common one. "Relation" as a general concept, is an abstraction of any physical, physical – logical, logical or natural dependency between two or more things. "Relation" makes "relevancy" between things.

In the related things, new properties emerge not existed before relation. With removing the relations, also the emerged properties are removed. For example, consider a few pieces of short ropes that may not be used in single. But when tied or twisted to each other, gain new useful properties (length or strength) that never were before. According to the dependency between things, there may be different types of relations among them.

"Relation", according to its nature, is definable between at least two elements. In studying relations among various elements, it is better to break it into simple single relations between two elements. In this case, "a relation between two elements" is as a "unary\ simple relation" and "various relations between two or more elements" is as a "multiple\ complex relation".

The researcher of a system, by studying the structure and relations among the structural elements in the system, may study the effects of relations in the behavior and discipline of the system. In reverse, the researcher, by studying the structure, behavior and discipline of system, may discover the relations among its elements. It must be added that, continuation of behavior and discipline in a system, leads to some kind of "correlation" among its structural elements. "Correlation" is a common property among the various related elements that defines the degree of "dependency" or "relevancy" among those related elements. Changes in behavior or discipline of a system, may lead to changes in the correlation.

From a researcher's view, multiplicity and diversity of elements and relations in a system, is a main factor for "simplicity" or "complexity" of the system. From this view, complexity is having numerous, different and unknown parts or elements and relations among the parts in a whole or system. Also, ambiguity of the elements and relations and the roles of them in system and having unknown and unlimited interactions with the environment may be considered as complexity.

Relations may have some general properties such as "strength" or "weakness" or may be "permanent" or "temporary" in time. Researcher may classify relations in different classes such as "strong relation" and "weak relation" or "permanent relation" and "temporary relation". Emphasizing on "strength" or "weakness" and "permanency" or "temporality" of relations among the elements in a system, may reflect the strength or weakness of the structure of system and long time or short time relations among the elements from the researcher's view. In this case, the supposed "structural balance" of the system may be evaluated by the use of the same words.

If according to the researcher's criteria, elements and structural relations in the system are strong, then the structure of the system also can be strong. In addition to this, if there is some kind of correlation among the elements, then the life time of system also can be evaluated long. From the researcher's view, the structure of a system may have a certain life long with young, middle and old age periods. In other words, system may pass the three known main stages of birth, live and death.

For example, consider the natural systems like animals or plants. In the process of life of these systems, at the beginning we are faced with weakness in structure, behavior and discipline. Then gradually and for a specific length of time, we have strengthening the structure, behavior and discipline. At the end, again gradually we are faced with weakness in structure, behavior and discipline and finally destruction of the structure. The length of life in systems depends on many complex factors (including environmental factors) and the strength of structure is only one of these factors (of course the important one).

A "permanent relation" between two elements in a system, may have its origin from the early structure of the system. Also, it may be caused by the change in the structure or a result of behavior and discipline in a period of life of system. When two or more elements are from a common origin or have similar properties, then they may have some kind of permanent relation. Also, according to the time of relations, a "temporary relation" is a short time dependency between two elements as the outcome of the behavior or discipline of the system. By ending the behavior or change in the discipline, it may disappear.

As we saw in the above, in the case of relations among the elements in systems, the length of time may lead to strengthening the structure or reinforcing the behavior and discipline of the system or in reverse weakening them. In the life time of some systems this is due to the process of life in these systems.

It must be noted that strength or weakness are completely relative concepts and are applicable in relation with the length of life in systems. When a relation is the same in all times, then it is the same that must be and talking about its strength or weakness is not a subject of interest unless it is compared with other known relation.

In the case of system structures, there may be some scales. Based on the above, some of the scales are about the "structure type" and some other about "structural order". Since "matter" is the main substance in most systems, some of the scales about the system structures are from material views.

For evaluation of the structures of systems, there are different and very general scales. Some of the general scales are: "strength against weakness", "short length of life against long length of life, young\ new against old\ obsolete" in the case of elements and relations among them and "structural order against structural disorder" from the view of spatial arrangement of elements in accordance with the supposed goals for the system.

As we noted above, for describing the "coming together of parts as a whole", in addition to "relation", some other words or idioms are used. All these words or idioms less or more have similar concepts. In the remaining part of this section, we will explain in short some of these words or idioms from the view of the subject of this book.

"Interrelation" is a process in that two or more things "depend" to each other to create a whole structure. The process can be initiated from the "in" or "out" of the things participating in the process. "Interrelation" causes some kind of structural "dependency" between two or more things to form the desired whole structure.

"Connection" is a kind of relation among the elements in that a certain thing may facilitate the relation. In "connection", between two relating things there may be a third thing as the facilitator of the relation. In this case, for example it is told that this thing is "connected" to that thing by this facilitator.

Also "connectivity" is the ability of connecting or the amount or degree of relation in the connection. Indeed, "connectivity" points to a logical dependency among the elements of a system or sub-systems of a main system. Also the "connexion" as "bringing two things together" is to create a kind of structural relation between two things for the purpose of "one to be with the other".

"Link" is a kind of relation in that by adding a thing, some thing is related with another thing. Also, link is a relation between two things in that one of the things affects the other thing. Some times, link may be viewed as some thing that relates one thing to another thing. Also, "concatenation" is a process in that the things are linked to each other as a chain.

"Contact" is establishing a kind of relation for a need or creating the appropriate background for establishing a permanent relation. "Contact" may be a relation for information interchange. In human relations, "contact" in some cases implies "coordination". In some other cases the word may be used for "conflict" in that some of the previous relations may "break". "Touch" is "contact" with "sense". Also "junction" is "coming together" and "conjunction" is "closing up together".

"Dependency" or "interdependency" is a relation in that some thing is affected or controlled directly or indirectly by another thing. "Dependency" or "interdependency" can be viewed as a unidirectional relation in that for example the first thing is "dependent" or "interdependent" to the second thing.

In the "dependent" or "interdependent" things, change in the first thing may cause change in the second thing. "Interdependency" is against the "independency". For example in the case of political relations, in the colony – colonialist relation, there is interdependency of colony to colonialist (for example from political views) from one side and interdependency of colonialist to colony (for example from natural resources in the colony needed by colonialist view) from the other side.

Interdependency also may be argued in a chain of linear steps. For example, production of a product in an industrial complex may be dependent to a chain of mid-productions in the same or other industrial complexes as mid-producers. In other words, "interdependency" is the dependency of two or more things in order to make a structure and especially to have the desired behavior or discipline.

"Fasten" is "bring together" and "make firm" a thing by another thing in order to protect it. Other words and idioms like "bind" and "tie" also more or less have the same meaning. "Juncture" is a thing for fastening. Also "associate" is "get together" and "association" is "getting together around a common purpose".

"Joining" is "going among". Going among implicitly implies that there are some elements in advance and now for meeting some requirements or because of new conditions, new elements go among the old ones. Joining may strengthen the former structure and in the case of disjoining, weaken it. The situation depends on the role of joined or disjoined elements. Also "joint" is "getting together" and "attach" means "joining a thing with other thing".

Systems Behavior

As we saw, "behavior" of a system is actions or reactions that system does against the internal or external events that some how affect the system. This leads to "stability" and achieving the "goals" of system. It must be added that in most cases behavior in a system is not a set of activities that system does it in a time for ever and then every thing finishes. In most cases, the concept of behavior is a set of activities some how periodic or repetitive. In some cases, the period or repetition has a certain understandable order and rule in space and time.

Systems with periodic or repetitive behavior with a certain order and rule, also display some kind of "stability" during behaving. This means that the total states or steps of the behavior, more or less remain the same in different periods or repetitions. While the details of the states or steps of the behavior are not exactly the same as before. This kind of systems may be affected by other systems, but always have the same total behavior. Usually, minor effects do not cause the fails in this total behavior. But major effects may cause the fail in the behavior or lead to completely different and new behavior by the system.

We saw that the systems with structure, when are subject to events, in addition to structural aspects (implying constancy and silence), also may have behavioral aspects (implying variation and motion while protecting and controlling the discipline). These systems usually may be the subject of interest by their behavior and discipline. Also the problems or difficulties of these systems usually are from their behavioral or disciplinal aspects.

It must be said that "constancy and variation" or "silence and motion" is one the big challenges man is faced with. "Behavior" in systems like the "verb" in sentences, for the sake of its nature and in relations with "constancy and variation" or "silence and motion", is the subject of interest or a challenging case in systems. In other words, "structures" and "disciplines" are important because of the importance of "behaviors".

While different events lead to different behaviors, also behaviors may cause other events and affect the system. This is continued until the system again reaches to its "structural balance", "behavioral equilibrium" and "disciplinal certainty" or "stability" in general. As a consequence, persistence of a change due to an event is related with the behavior from one side and with the structure and discipline from the other side.

In the process of studying the systems, trying to show that a set of events cause another set of events or presents a part of a certain behavioral pattern, is not sufficient. The reason is that, in the endless chain of causes and effects, each cause is the effect of another cause. For removing a difficulty, it can not be determined where must go in the back and what must be changed.

A behavioral pattern in a system is a method of functioning by the system in order to achieve certain goals or purposes. Instead of studying the endless chain of causes and effects, it is feasible to study the structure and discipline of the system. In this case, the ability for understanding the problems and finding solutions for changing or improving the behavior of the system will increase.

It must be said that, in most cases, the main causes of behavioral problems or difficulties in the systems are their structures or disciplines. While there are problems in structure or discipline, there will be problems in behavior. Continuation of this situation may lead to more complexity in problems. Preferably, finding solutions for the behavioral problems in systems must be started from going to the inside of the system.

The strength point of systems thinking is that, it presents certain "behavioral patterns" in the case of certain systems. Behavior of different systems, more or less obeys one or a combination of these patterns. Based on the dependency of behavior to structure and discipline, these behavioral patterns also dictate certain structures and disciplines.

In dealing with an unknown system in that only its behavioral aspects are apparent, with identifying the behavioral pattern of it, also the structure and discipline of the system will be identifiable. In consequence, by identifying the behavioral pattern, now the structure and discipline of the system can be studied and the causes of problems in the behavior can be found. By understanding the structure and discipline of the system and modifying it, the problems in the behavior of system can be resolved.

In order to study the behavior of a system, it must be identified that which common and known behavioral pattern matches with the structure and discipline of the system and can explain the current situation in the structure and behavior. Behavior in systems depends on time and implies a flow of time. In order to match the behavior of system with the common and known behavioral patterns, it must be identified that which parameters in system change with time. Then what is seen in the behavior can be seen in the parameters in some way. By studying these changing parameters, the behavior of system can be studied.

For a system, there may be different behavioral patterns. In order to study the behavior of a system, different behavioral patterns of it must be studied. It can be said that each set of events in a system, is related with a part of a "long term behavioral pattern" in it. Identifying the events, finding the relations among them and putting them in a set of related events, can help the researcher to abstract the patterns of the behavior of system from the sets of related events.

Since the processes that vary with time, usually are parts of the behavior of systems, one of the benefits of systems thinking in dealing the objects, their related events and the orders and rules in them, is the ability of it in dealing with changes in them with time. This approach is usually called "systems dynamics". In a general view, any aspect of behavior in a system can be in relation with an event. There are different methods and tools for making and displaying behavioral patterns in systems.

Behavior of a system from its researcher's view is a "chain of events, actions and reactions with constancy and variation or silence and motion caused by events", so that the results are important from the researcher's view. If this chain of actions and reactions is unknown or does not match with any known behavioral pattern, then understanding it is impossible or complex for the researcher. In the case of behavior, the purpose of researcher is to find patterns in this chain of actions and reactions and compare it with the known behavioral patterns. Then the researcher can judge about the system. "Behavior" is also as a set of "responses" to "requests" or related "events".

Systems have various behavioral patterns. Perhaps the most general and common patterns are as follows:
- "Continuous behavior": A behavior in that the different steps of system function are steady and not completely identifiable and distinct from each other. In consequence, the whole behavior of system is considered as a whole function.
- "Discrete behavior": A behavior in that the different steps of system function are in the form of different states and completely identifiable and distinct from each other. In consequence, the whole behavior of system is considered as sub-functions under a main function.

A "continuous" behavior of a system can be modeled as a "discrete" behavior and restricted to the states important from the researcher's view.

According to the above, in a general view, researcher can consider the behavior of the system of interest from "qualitative" or "quantitative" aspects. Mostly, "continuous behavior" tends to "qualitative" or "analog" aspects and "discrete behavior" tends to "quantitative" or "digital" aspects. In any case, events are the causes of continuous or discrete behaviors. Of course, as we saw, behaviors may cause other events and this chain may continue.

Events related with the behavior of system, may be internal or external to the system structure. In simple systems like a piece of stone in the previous example, "behavior" is any change like any movement of the stone caused by any external event. In complex systems like living beings, "behavior" is actions or reactions the system normally does against the internal or external events or in interactions with the environment.

In the daily conversations with natural languages, "behavior" usually appears in the form of "verb". It can be said that the "verb" in sentence, like the "behavior" in system, has a basic role in understanding the sentence by the listener or reader. Indeed, without the verb in sentence, there is nothing for understanding. This reflects the importance of behavior in systems.

Also in the discussions related with the behavior in systems, different words and idioms are used. These words and idioms while are different, imply some kind of "change" in systems. Among the all concepts in this case, "event" has a very basic and fundamental role. As we saw in the previous chapters, "event" is "any thing that occurs". Event implies change in the objects in space and time. Events usually are explained by describing their properties in space and time and related objects with their orders and rules.

From a researcher's view, some of the events may appear as "usual" or "regular" and some others as "unusual" or "irregular". Regularity or irregularity of an event means that the researcher can or can not understand it. Of course, any event for any researcher at first may appear as unusual or irregular. But by studying the various appearances of it, it may become usual or regular. Usual or regular events ordinarily are "iterative", "specified in space and time" and "ordered and similar". Unlike this, unusual or irregular events ordinarily are "non-iterative", "unspecified in space and time" and "disordered and dissimilar".

Indeed, in the case of iterative events, usually there are records and researcher can study the various appearances of the event in different spaces and times and extract the rule or understand the cause and effect relation of the event. But in the case of non-iterative events, usually there are not records and so extracting the rule or understanding the cause and effect relation of the event is impossible or difficult.

Usual or regular events are some how "predictable" and some times are as "normal" or "common and non-unique" events. But unusual or irregular events are "unpredictable" and some times are as "phenomenon" or "rare and unique" events. Of course, being iterative is not a reason for understandability and predictability of an event. There are many events occur one time for ever, but are understandable and predictable.

Also, while the details of the behavior of some big systems are not, but the total behavior of system is understandable and predictable. For example, predicting the weather in general and for a region (as space) and for days (as time) is easier than of predicting it for a point and for hours. So is the collective behavior when compared with individual behaviors.

In other words, knowing the rules of behavior of individual elements in a system does not help to predict the behavior of the total system. Although in the case of some systems the behavior is predictable for a long time in the future, but in some other systems it is predictable only for a short time. Instead of predicting the behavior of system for the future, it is better to study the behavior in reactions or responses to the events occur for system.

Evidences show that in the case of some systems, any change in behavior may be effective for a long time or even for ever in the future. But in some other systems, it is not so and the old behavior returns again after a time. It must be added that the behavior against an event, always is not the same that may be supposed.

For example in the case of narcotics and drug control, persisting on decreasing the supply of drugs, while the demand remains constant or even increases, is not effective. This policy increases the prices of drugs and consequently, may increase the crimes by the addicted drug users in order to get more money for the new higher prices. This is why besides the efforts to "decrease in supply", efforts are made also for "decrease in demand".

By finding the rule of an event, it can be concluded that if certain conditions or pre-requisites in the form of space and time are met, the event can occur again in the future. As we saw, events are the initiators of behavior in systems. Events in relation with the behaving objects or their origin (inside or environment of system), can be as "internal event" or "external event".

In general, "constancy" is the ability of protecting the existing structural, behavioral and disciplinal properties. In other words, "constancy" is remaining in the existing conditions by protecting the current situation and resisting against any change in the properties in time. Unlike this, "variation" is having flexibility in the structural, behavioral and disciplinal properties and the ability of changing the properties in time. Constancy is a criterion for silence and non-freedom (determinism) and variation is a criterion for motion and freedom (non-determinism) in the space and time.

As we said, one of the big challenges for human being is how to deal with "change". Indeed, most of the difficulties come from the change. Without change, there was not any problem and all the things were the same as they must to be. But the reality is not so and change is in the nature of many things. This is why the behavior is an important and central subject in systems.

Constancy and silence in a system means that the time is not important in it. But in variation and motion, time is important. Variation and motion or permanent changes, is the main feature of the world we live in. Supposing constancy and silence or permanent stability in its abstract concept in this world, is equal to inexistency.

All the evidences show that the time goes on without an end. In contrast, the reason for the flow of time is the changes we see in the world around us. Indeed, we experience the flow of time by seeing the changes or variations and motions in our world. Apparently, there is no end to this.

The degree of variations in a system is determined by its "change rate". "Change rate" can be evaluated as "high" or "low" by the researcher who studies the system. In this case, "low change rate" can be as sign of "rigidity\ hardness" and "high change rate" as sign of "liquidity\ softness" of the system from the structural and disciplinal views.

Most of the changes in systems come from their behavior. Of course, when a system behaves in an environment, also there may be changes in the environment. Any change in the environment of a system may facilitate or in reverse may cause difficulties in the behavior of the system. Systems when changing during behaving also may change due the changes in their environment. Also some of the changes are out of the ordinary behavior of the systems.

If a system of interest changes in some way, it must be understood what the change is and how the change occurs. If the change is out of the ordinary behavior of the system and has not managed before, how it can be managed. The difference between an efficient and inefficient system is that in an efficient system, unwanted changes are identified and managed at its right time. But in an inefficient system unwanted changes may be identified very late. In consequence, the structure or discipline of the system may damage or the system may collapse.

From a researcher's view, change in systems may cause transformations and this may be evaluated positive or negative. As we will see, based on the evidences, in a normal condition, the endless flow of time in systems leads to growth and evolution in one side and destruction and extinction in the other side. Going of "old" and coming of "new" is the manifestation of this trend.

From silence to motion, the importance of time increases. For this reason, the researcher in dealing with lower or higher rate of changes is forced to simulate the lower rate changes in higher speeds and in reverse, the higher rate changes in lower speeds. In studying the systems, researchers usually try to simplify the subject of study. To do this, they suffice to the cases with no or less change in time. Also in evaluation of the change, restrict it to the ranges that they can understand and manage it. In studying the systems, usually it is supposed that some of system properties remain constant or change in a specified range.

Any change depends on the existing conditions. If the range of this dependency supposed from "very high" to "very low", then there may be a question that in a specific case what is the amount of this dependency? A point may imply a full "dependency" with the meaning that "any future is fully depended on its related past". Other point may imply a full "freedom" with the meaning that "because of the changes in conditions, even minor changes, no future depends on its related past". The middle point may imply some type of "dependency and freedom" or "determinism and non-determinism" with the meaning that "although the past has effect on the future, but this effect is not fully deterministic".

From a researcher's view, any combination of "dependency" and "freedom" noted in the above may be applicable in studying the systems. Knowing the change pattern of systems is important in knowing and predicting what will occur in the future. Also, the analysis of change helps to know what currently is occurring.

"Change" can be discussed in the following categories. A system may experience all of these in a period of time:

- Steady Change: "Steady change" is a constant and continuous change in time. Of course all of these changes are not fully constant and continuous, also may be some how progressive. Steady changes with a certain pattern, can help to determine the length of time of continuation of the process that implies the change.

- Cyclic Change: "Cyclic change" is repetitive and discrete in time. This is caused by the effects resulted from "feedback". In cyclic change knowing the period of change and amplitude and time of maximum and minimum of change is important. Some of cyclic changes while are possible to occur in the future, but are not predictable.

- Chaotic Change: "Chaotic change" is a change without any order or rule and its space and time of occurrence is unknown. Mainly, "chaos" or "anarchy" is against the order and rule in space and time. Chaotic changes are mainly unpredictable. Of course most of the chaotic changes really are the unusual or irregular changes without any known order or rule at first. Some simple and known processes when repeated in different cycles, then may become complex and appear as a chaotic change.

"Cause and effect" is one of the oldest subjects in the behavior of systems. In dealing with an event and the "effect" of it, we usually try to find the "cause" of the "effect". "Cause" is the initiator of the event that has led to the "effect". "Effect" is the "result" of the producer of it as "cause". In other words, in the common rule of "cause and effect", any cause has an effect and any effect occurs by a cause.

In an ordinary sentence, when a noun is used as the subject of a verb, then the verb might occur on another noun as the object of the verb. In this case, the effect of the subject of the verb (as the cause) on the object of the verb (as the effect) can be identified by the meaning of the sentence. We can say that "cause" implies subject and "effect" implies object of a verb.

Cause and effect can be seen as a "process". "Process" is a flow of "inputs, functions, outputs". "Function" is a set of actions done on inputs to produce outputs. "Inputs" and "outputs" is any thing in the form of matter, energy and information entered to as raw or exited from as processed material. Transformation of inputs to outputs is done by the function of the process.

In the past, for man, effects were the deterministic results of the causes. Today, the effects are not necessarily deterministic, but are the results of the functions. If we change the functions, then the results also change. By changing the functions, we can have effects very close to one that we want to have.

The simplest form of cause an effect is that the flow that connects the cause to the effect is direct and linear and goes in one direction as a chain. This form of cause and effect may be used in many cases in every day life. This form is useful when there is enough certainty about the effect resulted by the cause.

But in many real cases, instead of a linear and unidirectional pass, we are faced with a closed and bidirectional progressive pass from cause to effect and in reverse. In other words, although any effect has a cause but the effect also affects the cause. In other words, flow from cause to effect instead of being linear and "open" is circular progressive and "closed" pass.

According to the above, in behavior with open pass, usually there is not any mechanism for control of behavior. But in behavior with closed pass, usually there is a mechanism for control of behavior. When behaving, if the structure also changes, then usually there is a mechanism for protecting the structure. Information comes to scene in closed pass behavior and plays its role in protecting the structure and controlling the behavior. As we saw, the open linear pass is not useful in the complex cases. It is useful only in simple cases or when the domain of research of the related event is limited.

In some cases, the cause may be the effect of another cause and this chain may continue in backward. In systems thinking, what restricts going back for finding the previous causes is the "environment" and "domain of definition" or "boundary" of the system. As we know, the linear chain of cause and effect does not consider all the aspects like the act of effect on cause. This is why it is not useful in systems thinking. In systems thinking, the cause and also the act of effect on cause in order to produce the desired effect in a closed pass, is desirable.

Circular closed pass of cause and effect is as a "process" of cause and effect with "feedback". "Feedback" means returning some part of output of a process again as input to it in order to control the process until the desired function or behavior is done properly. Indeed, "feedback" is a closed pass of action and the information related to it.

For example, the early wind or water mills also had a feedback mechanism. In the early mills, the rotation speed of the upper stone of the mill was appropriate with the intensity of flowing wind or water. In other words, less intensity of wind or water led to less and more intensity of wind or water led to more rotation of the upper stone of the mill. The feedback mechanism acted so that the decrease in the intensity of wind or water caused decrease in the slope of grain feeder and in consequence, less grains to enter to the mill. In contrast, increase in the intensity of wind or water caused increase in the slope of grain feeder and in consequence, more grains to enter to the mill. Grains entered to the space between the lower fixed and upper rotating stone of the mill through a hole on the upper stone.

In the "feedback" process during the behavior of system, changes in the amount of an important variable, indirectly affects the future amounts of it. In this closed loop, the amounts of this variable repeatedly evaluated and according to that, changes are made in the behavior of system towards the system goals. As we see, the behavioral pattern with feedback, with the aid of information, indeed relates with discipline in systems. "Feedback", as the closed pass or the non-linear flow of cause and effect is against the open pass or linear flow of cause and effect in the behavior of systems. Since the feedback is related with different behavioral patterns, it is very important in systems.

It must be added that for systems some kind of "life" can be imagined. Also "to continue the life" in systems implies the existence of a behavioral pattern with repetitive constancy and variation or silence and motion including feedback loops. Indeed, "life" is the "repetition" of these processes with constancy and variation or silence and motion. Considering the feedback loop helps to find out the cause of a specific behavioral pattern in a system. The causes of appearance of a behavioral pattern in a system usually can be studied inside its feedback loop mechanism.

For example, consider an every day life activity like filling a cup of coffee from a kettle. In this example, the first thing is the "coffee in the kettle" and the second thing is the "level of coffee in the cup". The "goal" of this activity is having a certain level of coffee in the cup, not overflow from it. When doing this work, the person repeatedly controls the level of coffee in the cup. Indeed, the feedback loop of control can be imagined by what the person does in this work. By pouring more coffee in the cup, with increase in the coffee from kettle, also there is increase in the level of coffee in the cup in the same direction.

Now, instead of "level of coffee in the cup", consider the "amount of free space in the cup". In this case, by pouring more coffee in the cup, while there is an increase in the coffee from kettle, also there is a decrease in empty space of the cup. The person, by evaluating the amount of decrease in the empty space of the cup, can control pouring of coffee from kettle to the cup. If instead of filling the cup, to empty a full cup of coffee until to have a desired level of coffee in it is considered, then the process is reversed.

Like the cause and effect relations, also there are many processes in the world around with a model of behavior called "stock and flow". Indeed, most of the processes in daily life or production and industry, behave in a way including stocks and flows of various things. For example, an ordinary water reserve or a bank account includes stock and flow of water or money. Also there are many other systems with this kind of behavior.

"Stock" is "adding on" of some thing and "flow" is the "movement of" that from one "stock" to another "stock" and finally "consume" it. "Stock and flow" includes an origination and a destination and the flow from origination to the destination with a specified "flow rate". Based on this concept, we are faced with things as "linked vessels" in the world around us. With this assumption, all the decreases in some places cause increases in other places. Also, all the increases in some places cause decreases in other places. In other words, most of the things are stocks or flows from one stock to another stock.

For example, when an employee receives salary from an employer, equal to the amount of salary, is subtracted from the account of employer and added to the account of employee. Employee uses salary to buy things. In this case, equal to the amount of any buying, is subtracted from the account of buyer and added to the account of seller. Also in the case of bought articles, equal to the amount of articles, subtracted from the shop or stock of seller and added to the consuming stock of the buyer. Finally, the seller must supply the sold articles and this process continues infinitely and permanently. The coffee and cup example discussed above, is also another example of stock and flow process.

Identifying "stocks" from "flows" is not a simple work. One way is to see what happens in the process when the time is stopped. If a quantity appears non zero and measurable then it is a "stock". But if the quantity is not measurable, then it is a "flow". This means that the "flow" is meaningful only in time. In other words, when the time stops, nothing can flow. Also the "stock" is related with "constancy" or "silence" and "flow" is related with "variation" or "motion".

If the very general types of stocks include "matter", "energy" and even "information" (but a little different), then from matter to energy and finally to information, also the "liquidity" of the stock increases. It is clear that the transfer or "flow" of matter is harder than energy and same as this, the transfer or "flow" of energy is harder than information. Energy is less concrete and more abstract than matter and it can be easily transferred to other point for example by electric wires, but it is not easy in the case of matter. Same as this, information is less concrete and more abstract than energy and it can be easily transferred to other point for example by talking aloud. Possibility of transfer or "flow" of these three main resources in organizations (usually between points), increase from matter to energy and finally to information.

As we saw, in an industrial complex, if the flow of energy is easier than matter, the flow of information is also easier than energy. This liquidity or indeed this "need to liquidity" of information that was not possible in the past but is possible and important today by the use of information technology, reflects the importance of information in the processes. Energy is as blood and information is as neural signals that circulate in the whole structure of the organizations. This is why that gathering, saving, processing and distributing of energy and information is a central activity in today organizations and electrical energy and information networks are so important in today societies.

"Action" in a system is a part of behavior that is done against the occurrence of an event in the system. In other words, action is related with event. "Action" can be done in system in all times. Also "reaction" in a system is a part of behavior that is done against an (usually external) "action". In other words, "reaction" is a "necessary action" against another "action" and is done after the action in time. Action is what is done by its doer and event is what is seen by its observer. Event also is a kind of action or initiator of action. For example, while "coming" and "going" are as actions from the doer's view, also are as events from the observer's view.

In a system, "request" is an action that implies a demand for some thing, doing some action or regarding some cases that a first side asks from a second side in a mutual relation. In contrast, "response" is a reaction that implies the answer or supply of the demand that the second side does against the "request" of the first side. Doing "reaction" against "action" naturally (by Newton's Law) is a must, but giving "response" against "request" is not.

When a system behaves, if the action and reaction between two elements is mutual, then there is "interaction" or "mutual interactivity" between the two elements. "Interactivity" is the give and get between two or more elements in order to achieve a desired goal. Also "intercourse" is a kind of give and get between system and environment or other system in order to protect the stability of the system.

"Function" of a system is the result of all actions and reactions of the system in order to create the conditions or things that define the goals, purposes and ideals of the system. In other words, "function" of a system is the set of activities in the form of "behavior" that are done in order to achieve the "goals". "Function" is the core of "process".

As we saw in the previous chapter, "process" is a chain of related actions with a certain goal. Process implies "getting inputs", "functioning on inputs to produce outputs" and finally "giving outputs". In a process, each step of action, takes us one step nearer to the goal. "Inputs" are as the primary raw resources (including raw matter, energy, and information) that by "functioning" on along the process are transformed to secondary processed resources (including processed matter, energy, and information) as "outputs".

In the case of processes in a system, it is needed to view the "whole" and "environment" or what is known as the "big picture" of the system from the out side. This leads to a wider view of the system. To do this, first the processes that do the core duties of the system and are called "core processes" must be identified. Then the processes around these core processes and their relation with core processes must be identified and rated in their positions.

Achieving the goals requires maintaining a stable state during the behavior of system. Stable state includes balance in structure, equilibrium in behavior and certainty in discipline. This state is affected by the "environment" or the pervasive system that includes the system of interest. Discipline in the structure and behavior of systems and the dependency of it on information is also arguable from this view. As we saw, information is a factor in determining the amount of orderliness in the structure and regularity in the behavior of systems. Since the behavior is more related with variation and structure is more related with constancy, behavior reflects more urgent information than structure.

Goal, purpose and ideal in systems are also meaningful in relation with behavior. Because, any system interested form behavioral view, from its researcher's view seeks certain goals, purposes and ideals. In other words, any system during its life time can seek certain goals, purposes and ideals. Systems acting in this way are called "goal\ purpose\ ideal seeking system". In contrast, being in a wrong way in that the system can not achieve its goals and purposes, some times is as a "trap" that releasing from it is not easy.

In a system, "goal" is a certain preferred state (like passing the elementary exams) that is achievable in a specified short time. In other words, "goal" is a favorable and desirable situation that a system with a certain structure by doing a certain behavior wants to achieve. "Goal seeking" in systems is behaving in a way to reach to that specified state or situation. This is done by the aid of some type of "memory" and the possibility of selection in behavior in similar or different situations with similar or different methods. From a researcher's view, a "whole" as a system, usually has "goals" related with different "purposes".

"Purpose" is a certain preferred goal and is achievable in longer time than goal. Indeed, "purpose" is a situation or step that the system must achieve certain goals before achieving it. "Purposes" are achievable by achieving a set of elementary "goals". In other words, achieving different stepwise goals (like passing the elementary exams or winning the regional tournaments) may be achieved in order to achieve a certain purpose (like passing the final exams or winning the national tournaments). "Purpose seeking" in systems is behaving when there are different goals with possibility of change in them but all with a certain purpose.

"Ideal" is a certain preferred purpose achievable in unlimited length of time. In other words, ideal is a purpose that achieving it is possible in theory but impossible in practice. By any struggle in this way, system becomes nearer but does not reach to it. "Ideal seeking" in systems is that system by reaching to goals under certain purposes also can follow other goals under other purposes.

"Dynamism" in general that includes "behavior", with positive or negative dimensions, can be imagined in three main forms as "growth", "destruction" and a combination of these two. "Growth" leads to "evolution" and "destruction" leads to "extinction". Evidences also show that the behavior of most of the dynamic systems some how obey these three basic and main patterns. These patterns can also be imagined in different forms. Based on these basic patterns in the behavior of systems, we can imagine different forms of behavior in systems. The forms of behavior differs from uniform linear to continuous growth or destruction or delay and change the direction to opposite side or oscillation form of growth and destruction.

"Growth" can be in the form of "linear growth" or "exponential growth". "Linear growth" is like "arithmetical progression" and "exponential growth" is like "geometrical progression". A growth process, because of the plentitude of necessary resources in the environment, may be fast in the beginning. But gradually and with decrease in the resources in environment, may slow down. Finally the system with combining these increase and decrease, may reach to a new "equilibrium" in its behavior.

Indeed, "destruction" is the "growth in reverse or opposite direction". Destruction like the growth also can be in the form of "linear" or "exponential". But, in destruction the situation is opposite to growth. A destruction process because of the resistances due to the structural protection or behavioral control may be slow in the beginning. But by the time and breaking up the resistances (also as a result of destruction), destruction also may be accelerated.

"Growth" and "destruction" because of the "restriction against growth" and "resistance against destruction" caused by the system or its "environment", have an oscillating form with an "oscillation amplitude". This situation can be seen in the nature. In other words, growth in a system can continue until there are necessary resources in the environment. Also destruction in a system can slow down due to natural or prior preparations in systems. This leads to a "wave form of growth and destruction" in the behavior of the systems.

For example, the growth of a population of living beings inhabited in a living environment and at the same time the resistance of them against death can be noted. As a result, in any way, systems tend to achieve to a "balance from matter view", "equilibrium from energy view" and "certainty from information view" or a "stable" state in general. In other words, according to the conditions, like the wet and dry periods of climate, growth and destruction can have a wave form with decrease and increase.

We saw that the "oscillation state" is a combination of growth and destruction inside the "oscillation amplitude" and system is stable in it. Hence the curve for displaying this oscillation has a wave or sinusoidal form. All of these are for conditions that the system maintains its "stable" state and continues to live. If the changes or variations are out of the oscillation amplitude implying the "stable" state, system finally may be destroyed.

"Evolution" is a growth with gradual progressive changes by passing from mid steps or states in that the structure and in consequence also the behavior and discipline of the system are improved and promoted. By evolution, an old form is transformed to a new form. The general theory of evolution has its roots from the ancient Greek. According to this theory, a current form and content has come from a past form and content. This is done by gradual progressive changes with less or more distinct methods. Evolution can be seen in many natural change processes. Like growth, the evolution also continues until the system has the ability and capacity of growth and evolution and there are necessary resources in the environment.

Unlike "evolution", "extinction" is destruction with gradual regressive changes by passing from mid steps or states in that the structure and in consequence also the behavior and discipline of the system are weakened or retrograded and finally the system is destroyed. The concept of extinction indeed includes the final state of destruction in that the system destroys. Like the evolution, also we can see the extinction in many natural change processes. Extinction may occur because of the obsolescence and especially the destructive changes in the environment of the system. Like destruction, extinction also occurs when the conditions and resources are not good and finally push the system to the final state of destruction. But if the conditions change, then the system under extinction may be rescued from the final destruction.

"Learning" is a process in that the ability to increase the efficiency and effectiveness in behavior to achieve a goal is improved. This is done by repeating the behavior to achieve the goal. Learning requires ability of remembering the past behavior and modifying it in achieving a goal. Learning is possible when there is the possibility of selecting a behavior among the various possible behavioral patterns in achieving a goal. By learning, systems can be adaptive and self-adaptability is a process in learning in systems.

"Adaptability" is the ability of changing the self or the environment in order to increase the efficiency of functionalities in systems. Adaptability may be done in the form of changing the internal conditions by the change in external conditions or changing the external conditions by the change in internal conditions of the system. Among the possible ways for changing the conditions in systems, changing the internal conditions is more common than changing the external conditions. An "adaptive system" is a system that when the efficiency or effectiveness of its behavior decreases, can change the internal or external (the environment) conditions. By doing this, system can improve the efficiency or effectiveness of its behavior in order to achieve the goals.

"State maintaining system" is the simplest of the adaptive systems. This kind of systems, by changing the external conditions can maintain their desired fixed state. But, the complex adaptive systems, when the external conditions change, may reach to some kind of compatibility with the environment. It must be added that "state maintaining system" is a system with a behavior to maintain its state. This is done as a reaction against the similar or different internal or external events. Most of the systems with the name ending to "stat" like "thermostat" are this kind of system.

To continue having structure and protecting the structural order and having behavior and controlling the behavioral rule in time, can be viewed as the "system life". By this view, any system as "life cycle" can "be structured or born" in one time, "behave or live" for a length of time and finally "go out or die" in another time. This may be viewed once for ever or in a repetitive form. In other words, there are different kinds of life cycles for systems. Like the cause and effect chains, system life cycles also can be in two major types as "linear\ start – stop life cycle" and "circular\ periodic life cycle".

Linear life cycle is a linear path consisting of "start – exist\ operate – stop" or like in living beings "birth – live – death". In circular life cycle this path is repeated periodically. The repetition process like "turning on - operation - turning off" in an ordinary electrical device, may be the same or like in plants may be evolutionary. In other words, start and stop repeated along the time. From system view, in the case of system life, instead of a linear or "open", usually we are faced with a circular or "closed" life cycle.

In the case of behavior, different scales can be defined. Since the behavior is the central point of attentions, when compared with other aspects in systems, more scales about the behavior can be defined. "Energy" is the main resource needed for behaving. As result, some of the scales are definable around the energy and some others around the function by the use of energy.

Some of the most general scales for the behavior of systems are as follows:
- "Efficiency": The ratio of the right acted cases to the total acted cases "without attention to the quality" of the cases in the form of system behavior.
- "Effectiveness": The ratio of the right acted cases to the total acted cases "with attention to the quality" of the cases in the form of system behavior.
- "Throughput": The ratio of the "output" to the "input" or the "real output" to the "expected output" from behavioral view.
- "Productivity": The ratio of the "outputs" to the "average outputs" or the "acted struggles" to the "expected struggles" in the form of system behavior.

- "Profitability": The ratio of the "consumed resources" to the "expected to consume resources" in the form of system behavior.
- "Flexibility": The ratio of the "real compatibility" to the "expected compatibility" in the form of system behavior.
- "Performance": A combination of all of the above mentioned scales for evaluating the behavior of a system. Indeed, "performance" is a total scale of "how" the system is and the "measure of performance" is a criterion for judging about achieving the goals in the system. The gathered data for this scale also may be used for modifying the behavior of the system. In relation with behavioral patterns, this scale also may be used in discussions about the function and behavior of the systems.

Systems Discipline

As we saw, "discipline" in a system is any king of "structural order" in "space" and "behavioral rule" in "time". Discipline also includes the mechanisms necessary for protecting the structural order and controlling the behavioral rule. This leads to "structural balance", "behavioral equilibrium" and "disciplinal certainty" and in general "stability" in the system behaving to achieve the "goals".

"Protecting the structural order" and "controlling the behavioral rule" are as resisting against the structure breaking and behavior disturbing changes during the behavior. For this, there are mechanisms in systems. Usually it is supposed that the desired behavior in systems is possible only when there is a certain order among the elements of system, in the form of its structure. While there is some kind of stability in the structure that we call it "structural balance", system can continue its behavior properly.

"Order" is a supposed spatial arrangement of the system elements that must be maintained in order the system to behave properly. In other words, although the system may take different apparent states in different times, but always has some kind of "structural balance". Also "control" is monitoring the behavior to be done properly according to its rule in time. In other words, in the behavior by the use of the available energy, "control" wants to establish some kind of "behavioral equilibrium".

The result is that the system does not deviate from the supposed path for its behavior for achieving the "goals". As a result, system can continue its behavior along the time in order to achieve the goals. In other words, a stable system is a system subject to events and with behavior due the events that while "protects" its structural order and "controls" its behavioral rule, can go towards its "goals" continuously.

"Protection and control" are among the important subjects in systems. As a result of this, system always maintains its proper structural order and behavioral rule. Also discussions about the "organization and management" (structural order and behavioral rule and protecting and controlling it), are among the important subjects in the world today. These subjects and their importance today, reflect that the systems in addition to structural and behavioral views as before, also are important from disciplinal views. Importance of information and its technology today, is due to this importance.

From disciplinal view, the requisite for maintaining the stability is continuously receiving the information related to the events during behavior and processing and using it for protecting the structural order and controlling the behavioral rule. As we see, "information" is also a necessary basic resource like "matter" and "energy" in systems. Without the availability of necessary information, the function of the system, important from its disciplinal view, will fail. In this discussion, "order" and "protecting" it implies some aspects of "space" and "rule" and "controlling" it some aspects of "time".

In order to have a simple imagination about the discipline in the world around us, suppose a car parking place with no line drawings for parking the cars. In this parking place, drivers will park the cars without any order or arrangement. In such a situation, no body will know the number of the places for parking the cars in the whole area of the parking place. If there are some cars parked in the parking place, also the number of the remained places for parking the new cars will be unknown.

Most of the parking places in the past were or some of them even now are such places. Based on the subjects discussed in this book, we can say that, this type of parking places has only structural aspect or "place to park cars". In other words, in this type of parking places, there is not any "order" in the case of park places of the cars or any "rule" in the case of parking the cars in the proper places by the car drivers.

Now suppose that this parking place has line drawings for parking the cars. In this case with a look to this place, at least explicitly any body will "find" that there are some or are not any places for parking the new cars. If there are places for parking, then approximately or exactly, the number of empty places for new cars will be known. In this case, we can say that now the parking place has some kind of "structural order". The requisite for "protecting" this order is to park the cars in the proper places designed for a car.

Now suppose that enter and exit passes also have line drawings and a man as the parking charge or responsible of the parking place, stands in enter and exit point. Also suppose that the man has a number of cards in hand equal to the number of park places for cars, each card for one car. In this case, the man when opens the parking place for daily services, can give a card as "receipt" to the driver of any entering car and get again the card as "voucher" from the driver of any exiting car. It is clear that the man has cards in hand if there are places for new cars in the parking place. In consequence, the man accepts new cars while has cards in hand.

Now we can say that, this parking place in addition to "structural order" has some kind of "behavioral rule". The requisite for "protecting" this "order" and "controlling" this "rule", is parking the cars in the proper places by the drivers and monitoring this by the man as the parking charge or responsible. If this parking acting like this for a long time and by the time the damaged line drawings and cards are maintained or renewed and always the structural order is protected and behavioral rule is controlled, then we can say this "order and rule" are "settled down" and every one as the parking responsible or car driver will obey it.

Now suppose that the cards are numbered. In this case, the man, when giving the cards to drivers, can write the numbers of card and car and the date and time of entrance in a raw of a simple table as a list. Also suppose that the man, when a car exits from the parking, first checks the numbers of card and car with the numbers in the list and writes the date and time of exit in other columns in the same raw of the table. In this case the man can calculate the parking charge and get the money from the driver.

The man may think that to number the car park places like the cards. In this case, the man can ask the drivers to park the cars in the places with the number on the cards. In entering or exiting, drivers directly and easily can go to their places that are arranged by numbers on the cards. It is very helpful especially when the parking is big and a person other the first driver comes to take the car.

With the works mentioned above, parking will find a proper "organization and management" situation. Now add the guiding signs related to the numbered places, magnetic cards, computers and the related information systems to this. This indeed is the story of the evolution of the car parking places we have lived with them in our life time.

Now suppose that the numbers on the cars can be read electronically by a device when entering or exiting the parking place and a specific empty car park place with a code is assigned for every car entering the parking place. Also suppose electronic sensors that guide the drivers to park the cars in the assigned place. This is the current story of car parking places evolution in the information age.

In the case of order and rule in the structure and behavior of systems, there may be many classifications. The following classification may be one of most primitive ones:

- "Primary fundamental order and rule": The natural or inherent structural order and behavioral rule embedded or institutionalized or the basic information or order and rule in the system, like what is seen in the genes of living beings even before birth and interacting with the environment. This type of order and rule, reminds the "factory settings" in the case of technological systems.
- "Secondary defined order and rule": The acquisitive or definable structural order and behavioral rule or acquisitive or definable information or order and rule in the system, like what is seen in the living beings after birth and in interacting with the environment or in the case of human beings, trainings before entering to the work market. This type of order and rule, reminds the "initial settings" in the case of technological systems before staring to use the system.

From a system researcher's view, each of the above mentioned orders and rules can be "atomic" or "combinational" and "explicit" or "implicit". According to this, some may be "simple" and some others, "complex". From the system researcher's view, first or simple cases are somewhat "identified and specified" and the second or complex cases are somewhat "unidentified and unspecified". Also understanding the "combinational and implicit" orders and rules will need more struggle (energy) and certainty (information) than the "atomic and explicit" orders and rules.

From systems only with structural aspect (like the piece of stone in our previous example) to systems with structural, behavioral and disciplinal aspects (like the living beings), "simplicity" decreases and "complexity" increases. The natural tendency of the system researcher's is that the orders and rules of the structures and behaviors of the systems to be atomic and explicit or simple. As result, one of the cases that a researcher may encounter with is moving from "implicitness to explicitness", "combinational to atomic" and "complexity to simplicity" in the form of a "model", that may be a big challenge.

As we saw, "structural order" in a system is any logical and understandable arrangement of the "elements" of system in "space", in order to achieve the desired "wholeness". Order implies that the elements for achieving the wholeness, all are properly placed in their spaces. Order finds its meaning by the "relations" among the elements and their spatial arrangement. Order, a part from its formal appearance, is an emergent property of the wholeness. Of course, some times, structural orders have specific formal appearances. Evidences show that some of the patterns in the nature, display specific meaningful appearances of order.

For example, "equality", "similarity" and "symmetry" are among the simple but common patterns in the nature or man made systems that display certain appearances of order. "Equality" and "similarity" at least need two things. But symmetry usually can be seen in one or more things and at least in two or more aspects or parts of some thing.

"Equality" is a situation in that two or more things (like breaks in a building) considered being the same. In equality as a structural property, for example, one part of the structure considered the same as the other part of it. Equality mostly implies quantitative aspects instead of qualitative aspects in the things. "Equality" is some how against the "diversity". Equality is also an order in that two or more things besides each other, from the view of structural properties in the things, all considered the same. In equality, being the same from structural and structural properties view is the subject of discussions.

"Similarity" is an order usually considered about the resemblance of one or more aspects of the structure of two or more things. In similarity, for example, one part of the structure is considered alike with the other part, from some aspects. Unlike equality, in similarity "being the same" is not the case, but "being alike" is important. In other words, in similarity the sizes are not important. Instead, whatever the ratio of the peer to peer sizes in the two things is the same, similarity is also more and more. In consequence, the two things are similar with higher degree of resemblance.

"Symmetry" is an order in that the structure is divided into equal parts around a separator as the "symmetry axis". This situation my attained by rotating the structure around an imaginary axis passing from the center of the structure. Symmetry at least includes two similar parts that against each other make the whole structure. Symmetry can be defined around one or more axis. Symmetry around different axis, lead to different structural shapes.

For example, the body of most living beings has a kind of symmetry. In this symmetry, by dividing the body as structure into two equal parts around an imaginary axis as separator, two sides become equal. This symmetry is some kind of "radial symmetry" that is called "two sided symmetry" or "bilateral symmetry". Also a rectangle or an ellipse has two (horizontal and vertical) axes or "four sided symmetry". A circle has infinite number of axes or "all sided symmetry". Other geometrical equal sided shapes like equilateral triangle and others, have three or more axes of symmetry.

As we told above, in some cases, symmetry may be defined around an axis passing from the center of the structure. This kind of symmetry can be seen in equilateral triangle or pentagon. In this symmetry, with some structural rotations around an imaginary axis passing from the center of the structure, structural equality around the axis remains the same. Symmetry around infinite number of axes or all sided symmetry is called "prefect symmetry". This kind of symmetry can be seen in sphere. In prefect symmetry, by any structural division or rotation around any axis, structural equality always remains the same.

"Homogeneity" is a kind of structural order with "structural uniformity" or simultaneous equality, similarity and symmetry. Homogeneity in its most complete form, with highest degree of equality, similarity and symmetry like in an ordinary sphere, implies that the parts of the structure as a whole, all take part in the same way in the whole structure.

Other concept of homogeneity may be that all of the parts of a whole are of the same type. When a researcher considers a whole as a homogenous thing, its concept may be that from the researcher's view, all the parts of the whole have the same structural properties. It must be added that from a researcher's view, homogeneity can take different degrees of structural uniformity. This degree may differ from one researcher to the other. Homogeneity can be considered only from aspects that the researcher is interested in.

"Arrangement" is an order in that the consisting parts of a whole are besides each other in the form of "first this, then that". Arrangement reflects a specific concept from the view of wholeness of the related parts. When the place of the constituent parts in the arrangement is changed, then the arrangement may be destroyed, still remain true or a new arrangement may appear. Also "combination" is an order in that the consisting parts of a whole are besides each other in the whole in the form of small groups without any arrangement in the groups. Now if any arrangement considered in the small groups, then we have an order as "permutation".

A case of the relation of mathematics with the orders in structures in nature is the Italian mathematician Fibonacci number series and its related golden ratio. Fibonacci number series is a sequence of numbers in that the first two numbers are 0 and 1 and the other numbers in the sequence are the sum of last two numbers (as 0, 1, 1, 2, 3, 5, 8, 13 ...). The property of this number series is that the ratio of the last number to the number before last number tends to 1.618 that is called the golden ratio. This ratio contains an order or ratio that can be seen in most of the structures in nature.

To imagine this ratio, it can said that dividing a segment of line into two unequal segments is with golden ration when the ratio of the length of the whole line to the length of longer segment is equal to the ratio of the length of longer segment to the length of shorter segment. This ratio is not only in geometrical shapes, but also in music and nature. Golden ratio presents some kind of order or structural balance that is used in art and especially in painting and music. In a painting, most of the important elements like the width and length, unconsciously and only by the sense of beauty by the artist, are selected or divided by golden ratio.

It must be said that the desired order in the structure of systems is not only from aesthetic views. Of course, in some eras in the history of man kind on the earth, in the case of some structures, perhaps some aesthetic views have been dominant. Remained old buildings with their especial architectures, abstracted for us today or even in the past only from their architectural views or "structural order", can be a confirmation for this. Some evidences show that in some cases, by anyway, some aspects of the visual appearance and the displayable order by the structures have been important in the past or even are today. In the discussion of "semiotic" (a structural view) and "semantic" (a behavioral view), this mostly tends to semiotics. Many cases of this kind of order like the especial architecture of the temples, churches or mosques can be pointed.

From systems view, "structural order" must be logically definable and understandable in relation with the wholeness that is called system. In order to have the desired behavior for achieving the goals in the system, this order "must" be in the structure of the system and "must" be protected as the "structure protection". Otherwise, structure of the system will collapse and its behavior and discipline will face difficulty and achieving the desired goals in the system will be impossible. As we saw, "structural order" as the spatial arrangement of the parts to make the whole as system and protecting it when the system is behaving, is essential for protecting the "balance" in the structure and finally "stability" in the whole system.

Of course, like in mathematics and especially what is discussed in "topology" as a branch of mathematics, even changing the appearance of some thing, may not change some of its properties. Also here, from systems view, the case is not protecting the "apparent view" of the thing as system, but is protecting the "wholeness" of the system. By this, system can continue its "behavior" in order to achieve the "goals".

As there is order in the patterns, also there is disorder in the nature. In the previous chapters we considered "entropy" form material view as a criterion for lack or existence of material concreteness in the structure of systems. We saw that from "structural order" view, entropy in one side is as a criterion or scale for "weakness" and in the other side is as a criterion or scale for "strength ness" in the structure of all systems.

Unconventional behavior in a system, like the scenes in western films in that the structure of a moving chariot in a stony road and out of control of charioteer, breaks down, can lead to "weakness in structure and losing the structural relations" in the system. The result is the increase of disorder or entropy in the structure of related system. Maintaining balance in structure and equilibrium in behavior by "protecting the structural order" and "controlling the behavioral rule" can prevent the increase of entropy. "Structure protection and behavior control mechanism" has this duty in systems. The result is remaining constant or under some conditions, decrease of entropy in open systems.

"Behavioral rule" in systems is the set of actions and selections that system does or acts against the events, in order to continue the behavior and to achieve the goals, while protecting the stability in the whole. We saw that behavior is accompanied with events and performable actions or selectable options in dealing with the events. Events are meaningful from the view of occurrence and priority or concurrency in time. Behavioral rule in systems also implies time in any way.

When from a researcher's view, system of interest has disciplinal aspects with the ability of protecting structure and controlling behavior, then the researcher may be interested in to identify the related mechanism for protecting the structure and controlling the behavior. It must be noted that "control" is the core of the behavioral rule of systems. Control is important because when the system is behaving, some of the properties of system (at least properties related with time) change and control wants to administrate or manage the changes.

"Behavior control" in systems includes the ability of identifying the related events and evaluating the situation or state of the system due to the events. This is done by comparing the occurred with the desired or predicted situations or states. This work is continued by selecting the desired cases related with the occurred situation or state and affecting the behavior according to the selections. As result, system can continue its behavior.

Control is done by continuously or periodically identifying and evaluating the events related to the various actions, in time. Control as a process, includes a closed loop of getting the data related to the current situation or state as input, evaluating it as processing and sending commands in order to adjust or modify the structure and behavior as output. Whatever the system is vital or near to organic systems, then the speed of repetition of this closed loop is faster and finally continuous that is a sign of being "alive". Behavior control in systems can be done in different forms.

In systems with behavior in the form of a process consisting of "input", "function" and "output", using some parts of "input" or "output" for controlling the behavior is a common control mechanism. In this type of systems if some part of system "input" is fed to it only for control purposes, then it is told that the system has a "feed-forward control" mechanism. Also if some part of system "output" is fed again as input to it only for control purposes, then it is told that the system has a "feed-back control" mechanism. Based on this, control mechanisms in systems can be in the form of "feed-forward" or "feed-back" or a combination of these two. In feed-forward control mechanisms there is some kind of "forecasting and preventing" and in feed-back control mechanisms there is some kind of "facing and avoiding" the unwanted situations.

"Feed-forwarding" is feeding especial inputs to the system in order "to adjust the structure and behavior" when the system behaves to achieve its goals. The result is "prevention" of "errors" in structure and behavior and finally "failure" in the function of system. This type of control is some kind of "monitoring" or "passive control". Also "feed-backing" is feeding some selected outputs again to the system in order "to modify the structure and behavior" when the system behaves to achieve its goals. The result is "avoidance" of "failure" in the function of system, because of the occurred "errors" in structure and behavior, when the system behaves. This type of control is some kind of "inspection" or "active control". According to the nature of the two control types mentioned above, it can be said that "feed-forwarding" is mostly useful for protecting the structural order and "feed-backing" is mostly useful for controlling the behavioral rule of the system when it behaves. According to the type of system, this protection and control can be done manually or automatic.

We saw that systems can maintain their structural balance and behavioral equilibrium and finally their stability, by doing the required adjustments or modifications on the structure and behavior, based on the protecting or controlling cases in them. As a result, systems can continue their behavior in order to achieve their goals. In artificial or man made systems, in practice, the control mechanisms of "monitoring" type, can be active in the form of "maintenance" activities. Also the control mechanisms of the "inspection" type can be active in the form of "repair" activities.

A system is transcendental if "monitoring" or "maintenance" is more dominant than "inspection" or "repair" in it. So that finally, instead of "inspection" or "repair", mostly there is "monitoring" or "maintenance". In other words, in theory, a system is supreme if has such a structural strength and concreteness and behavioral fluidity and flexibility so that always can continue to behave only by preventing the structural errors and behavioral failures.

For example, compare the process of maintenance and repair in an aerial and a ground transportation vehicle such as airplane and ordinary car, as two types of systems. Since the least error or failure in the structure or behavior of an airplane may cause the plane crash, there is no way except preventing it. Preventing the errors or failures is done as the "preventive maintenance" and by using the components with specified flying life times. After using the specified flying life time, components are replaced with new ones even if there is not any problem or difficulty in the component. But in an ordinary car, there is not such sensitivity and what is usually done, is the repair or replacement of the components if they fail to function properly.

Levels of feed-forward or feed-back controls may exist in some systems in the form of "internal control" (control mechanism inside of system) or "external control" (control mechanism outside of system). In some natural or artificial systems, internal control mechanism may be called as "embedded\ built-in\ natural control". In natural systems, some of the internal control mechanisms may be existed from the origins and some other may be created gradually in the life time of system by interacting with the environment. In artificial systems, internal control mechanisms may be embedded in the system in the factory during making the system.

Modifying behavior in a system, that may be possible by modifying the structure, needs some kind of "memory" and "learning" mechanism. In this case, outcome of control is the "gradual compatibility with the environment". As a result, system can maintain its structural balance, behavioral equilibrium and disciplinal certainty or stability in general and continue its existence. It must be added that control process usually leads to increase or decrease of some thing in the system. From this view, control can be "positive\ additive" or "negative\ subtractive". From a general view, control leads to protecting, maintaining or increasing order in the structure, rule in the behavior and certainty in the discipline. As a result, entropy remains constant or decreases in the system.

As we know, one of the properties of systems is the relations and interactions among its elements. When an element related with another element changes, then another element also may change in some way. Like this, with the change in the second element, required actions are done in order to affect the first element according to this change. This process continues until a new stable state is achieved in the system. Indeed, the protection and control mechanism is responsible for returning the system to its stable state.

Control of a system, instead of using the desired inputs or outputs for control cases, may be acted by the "internal clock/ timer" of the system. For example, occurring of certain events in certain times by the timer mechanism in systems (like washing machines) can be noted. Timer mechanism in system, can cause to transition of the system from a primary to a secondary step or state or initiate or terminate certain action.

For the discipline in systems, different scales can be defined. From the view of the subjects discussed in this book, discipline in systems contains two certain aspects including "structural order" and "behavioral rule". As result, some of the scales can be defined from the order in the structure view and some others from the rule in the behavior view. In the daily language, in the case of order there are interpretations like "strong order" or "irony order". Also in the case of rule there are interpretations like "exact rule", "stoned or stone written rule". Some times the purpose of "irony or stony order and rule" is the inflexibility in the related order and rule and in consequence, inflexibility in the related system.

"Order" according to its structural nature in this book, in a general evaluation, can be evaluated as "strong" or "weak". A strong order is not only concrete, but also is flexible and when the system behaves normally, does not break simply and consequently does not lead to readjusting the structure. In contrast, a weak order is not so and breaks simply even when the system behaves normally and consequently leads to readjusting the structure. Also in a general evaluation, "rule" can be evaluated as "exact" or "ambiguous". An exact rule is distinct and clear by itself and in similar cases does not lead to misinterpretations. In contrast, an ambiguous rule is not so and in similar cases leads to misinterpretations or out of rule decisions.

Desired order and rule in systems, is a strong order in the structure and exact rule in the behavior with flexibility in both. This property not only may be a reason for continued balance in the structure and equilibrium in the behavior of the system, but also will be desirable from the researcher's view. Normally, from the researcher's view, strong order and exact rule will avoid misinterpretations that may lead to the complexity of the case. Of course for researchers, understanding the strong order and exact rule may need more time and struggle.

Stability as Structural Balance, Behavioral Equilibrium and Disciplinal Certainty

We saw that the necessity of continuing the existence of a system in general, is the existence of "stability" in it. Stability implies resisting against the events that may lead to structural disorder, behavioral misrule and disciplinal uncertainty in the system. Stability from structural view implies some kind of "structural order" or "balance" from matter view, from behavioral view implies some kind of "behavioral rule" or "equilibrium" from energy view and from disciplinal view implies some kind of "structure protection and behavior control mechanism" or "certainty" from information view.

In a general sense and natural form, all the systems tend to achieve some how to a certain state as their stable state. After achieving the stable state, usually this state is maintained by the system until an unwanted event occurs and disrupts the current stable state. In this case, also usually after a time, a new stable state settles in the system. In other words, systems have some kind of feed-forward and feed-back mechanism that helps systems to maintain stability in a certain range of changes.

For example, an ordinary person as a system has the ability of maintaining the stability during the daily behavior. In the form of feed-forward the person is aware about the occurrence of some wanted or unwanted events beforehand and can deal with them properly. Also in the case of unexpected events, in the form of feed-back is able to deal with. A thermostat on a cooling or heating system or the mechanism for maintaining the body temperature around 37 degrees in centigrade, are other examples in this case. It must be added that more transcendental the system, more vital the need to information. In other words, needs to basic resources in order to maintain stability, is important from matter then to energy and finally to information.

For maintaining stability in systems, need to information is more urgent than energy and matter. Because in systems (like in human body), required energy and matter can be supplied and stored before needing it. But during the system behavior, exact information about the current state of behavior is immediately needed. Because, about the states in the future, exact information can not be gathered beforehand. Of course forecasting the future states in the behavior may be possible.

From the view of this book, "balance" is having some type of strength and equality from the structural view in different situations. Indeed, "balance" is the equivalency of the participated "masses" in the structure in all situations. "Structural balance" is one the three main factors effective in the "stability" in systems. Systems with the structural aspect, can be supposed to be in "balance" in all the life time of the system, unless an event may change the situation. Because the researchers of this kind of systems only are interested in the structure of the system and behavior and discipline is not the subject of their studies.

In a normal condition, systems with the structural aspect, after the end of the effects of an event, again may return to their balance state from the researcher's view. For example a piece of stone some where in the nature, always can be supposed to be in its balance state, even if kicked by a passer to another place. It can be said that in this kind of systems, "entropy" can remain constant, because from the researcher's view, they may remain in the same state.

In systems with the structural and behavioral aspects from the researcher's view, in addition to structural "balance", the behavioral "equilibrium" also becomes important. In other words, in this kind of systems, "stability" has two components as "structural balance" and "behavioral equilibrium". "Equilibrium" is to reach to some kind of continuity and permanency in the behavior in various situations. Indeed, "equilibrium" is the equivalency of the participated "forces" in the behavior in all situations. "Behavioral equilibrium" is also one of the three main factors in maintaining the "stability" in systems.

In other words, in these systems, while there is sufficient energy for behaving, in addition to "balance" in the structure, also there may be some kind of "equilibrium" in the behavior. In this case, "entropy" can remain constant. By cutting the required energy for behaving and as a result occurring difficulties in the behavioral equilibrium, also "entropy" begins to increase. Of course, in a normal condition, system again reaches to a new balance in its structure and especially new equilibrium in its behavior. This process is in the nature of most systems and the life of systems is a chain of these ups and downs in the structure and behavior.

For example, consider some warm water. If there is enough energy to keep the water warm, then the water can still remain warm. But if there is not enough energy, then the temperature of the water decreases gradually and finally equals to the temperature in the environment. In such a condition, the molecular motions of the water or the acting forces and the change processes that can do a work like in steam engines, come to stop. This situation continues until a new event occurs for the system. The water may become warmer or cooler than that is. Also in this case, after a time, system comes to a new equilibrium state again.

The idea of balance and equilibrium is useful in the systems with changes one after another. For example, we can say that the work market can be in its balance and equilibrium state, if the total number of the workless persons remains constant, even some persons may lose their jobs or find new jobs. But if the number of workless persons increases or decreases then the balance and equilibrium of the market also will disturb. Then gradually a new balance and equilibrium, for example in the form of decrease in the wages (because of the more workless persons) or increase in the wages (because of the less workless persons) will emerge in the work market.

Also it can be said that an ecosystem is in its balance and equilibrium state, if the birth and death rates of living beings in it are equal. But if the birth and death rates of the living beings are not equal, then the balance and equilibrium of the ecosystem will disturb. Then gradually a new balance and equilibrium, for example in the form of increase or decrease in the number some of living beings will emerge in the ecosystem.

Also if the supply, quality, price and demand for a product remain constant, then the market of that product can remain in its balance and equilibrium state. But if the quality of this product increases or decreases and the price remains the same as before, then increase or decrease in the demand, will push the price to a higher or lower level and fix it in the new level. In this case, also the number of the customers of this product will fix and as a result the market of the product will achieve a new balance and equilibrium state.

Now consider a system that from its researcher's view, all the three aspects of structure, behavior and discipline are important. This kind of systems, while the above discussion is true about them, can protect their structural order and control their behavioral rule during the behavior. This is done by the protection and control mechanism and the flow of information to system. The result is to achieve some kind of "certainty" from disciplinal view or protecting the order in the structure and controlling the rule in the behavior of the system. Cutting the information flow to the system, will lead to "doubt" or "uncertainty" and finally irregularity or stop in the behavior of the system. In this situation, if the system continues to behave, discipline of the system will disturb and system will face with disorder in the structure and misrule in the behavior and the result of it is "crisis".

For example consider an airplane with a safe structure and enough energy for flying as its behavior. During the flying, the airplane always receives the necessary information about the height, weather, speed of wind, obstacles ahead and other from outside of the system. The airplane also for achieving the "certainty" of being in the right route towards its destination, always is in information interchange with the air traffic control centers on the earth. In consequence, the airplane can protect its structural balance, behavioral equilibrium and disciplinal certainty or in one word only, the "stability" and continue to fly.

Continuing behavior by the airplane or to fly towards the destination in a normal condition depends on the safety of airplane, availability of enough energy and information interchange with the environment and air traffic control centers on the earth. Now if some of these cases face with some difficulties, then the flying of airplane also will face with difficulty. In this case, the structural balance, behavioral equilibrium and disciplinal certainty of the airplane may disturb and lead to a plane crash. As we see, in these cases, the importance of information is not less than energy and matter and the lack of information also can lead to difficulty in the structure, behavior or discipline of the airplane.

4

Systems Thinking

Introduction

"Evolution theory" led to the identification of more biological types and "periodical table of elements" identified other unknown or undiscovered chemical elements. In a similar way, the concept of system and systems thinking as a pattern and method can help to identify the unidentified dimensions of the objects and their related events. System and systems thinking and related concepts, can be used in discussing about the general similarities among the objects and their related events, approximately in all fields of human knowledge.

System and systems thinking and related concepts also are as a framework and method to approach and think about the objects and events of interest in the world around us. In other words, structural properties and orders, behavioral patterns and rules discussed in the field of system and systems thinking, can include all the objects and events in the world around and as a general framework or method can be used to identify all the objects and events. A certain structure, behavior and discipline or a certain set of structural properties, behavioral patterns and structure protection and behavior control mechanisms, identify a certain kind of systems.

Systems thinking implies that any object in general, from structural, behavioral or disciplinal view in alone or in combination with each other from the researcher's view, can be viewed as a system and studied by the aid of system and systems thinking concepts. System and systems thinking concepts also help the experts in the various scientific fields, to discuss their ideas or views about the objects and related events of interest, in a general format. This situation is due to the fact that the scientists or researchers in the various scientific fields, each some how are faced with objects and their related events in the form of a structure and structural order, behavior and behavioral rule and mechanisms for protecting and controlling this order and rule.

As we saw in the previous chapters, systems are consisted of or involved with the three main resources "matter", "energy" and "information". All the systems in a natural trend, in the form of "balance in the structure from matter view", "equilibrium in the behavior from energy view" and "certainty in the discipline from information view" tend to "stability" in general. Also, the "growth" and its consequence "evolution" and "destruction" and its consequence "extinction" are in the nature of most systems. "Growth" implies continuously using the three main resources for improving the structure, behavior and discipline of the system in order to achieve the goals. "Destruction" indeed is the result of decrease in the needed resources, increasing disorder in the structure and misrule in the behavior and collapsing the mechanism of protection and control in the system.

In a natural condition, each "growth" has the potential of "destruction" in itself. For example, the excessive growth of some living beings in an echo system with limited resources, by decreasing the resources in the echo system, may lead to the destruction of the same beings. It must be said that "constancy" and "variation" in "space" and "time" are the two big challenges most system face with.

As we saw, "constancy" includes silence and quietness and "variation" includes motion and diversity. In systems, after "constancy" usually there is "variation" and after "variation" usually there is "constancy" and this chain of constancy and variation continues. Evidences show that after any change there must be a fixation in order to say that the goal of the change has been achieved. Also after any fixation there must be a change in order to achieve the next goals and this chain has no end.

Unlike what it seems, we are not going forward in a linear unidirectional path with constancy and variation in time, instead of this we are faced with a circular progressive path. The linear unidirectional view implies that any problem due to the change, leads to an action as its solution and the path continues forward in this way. According to this view, any problem has a static state and for solving the problem, it is sufficient only to do a certain action.

But from the circular progressive view, any problem has a dynamic state. According to this, any action not only is based on the current conditions, but also affects the current conditions and the base of the next actions is the new ruling conditions. This process has no start or stop point and all the actions and interactions affect each other. Circular progressive path is an oscillating path with sequential ups and downs, constancies and variations or growths and destructions. Systems thinking help us to mange the change to reach to a fixation in order to achieve the goals and continue the process forward in this way.

As we know, according to the system concepts, in the world there is not any separated and isolated object in its ultimate concept and any thing in some way is related with some other things. But from a researcher's view, in some cases some of the relations may or may not be important. Determining the importance of relation of some thing with another thing depends on the researcher's view and the field of research. Based on this, in the world we live in, any object as a whole has certain properties and is consisted of parts related with each other in some way.

In a general statement, any object of interest to study or doing research on, not only is a whole consisted of related parts, but also is a part of a bigger whole and finally the world as the biggest whole. The idea that any object of interest while is a whole, also is a part of a bigger whole, implies that any object, indeed is meaningful in a bed that is called the "environment" of the object and is studied in this bed.

As we told, from a researcher's view, only the structure, only the behavior or only the discipline or a combination of these three may be considered as a system and the subject of research. But, by any way, any object according to its nature, from its researcher's view can be considered as a system in one of the three elementary, mid or high level as mentioned in the following.

Elementary level includes structure and structural order, mid level includes structure and structural order plus behavior and behavioral rule and high level includes structure and structural order, behavior and behavioral rule plus structural order protection and behavioral rule control mechanism in the object of interest. In this leveling of systems, "behavior" is the central subject and the systems are important because of the importance of their behavior. It must be added that the elementary level has no behavior but mid level mostly tends to linear unidirectional and high level mostly tends to circular progressive path. The tendency is because of the ignoring or not ignoring the disciplinal aspect in the systems.

"Systems thinking" as the "holistic" approach in the case of objects, indeed considered to be the opposite to the former "analytic" or "mechanistic" approach. In the mechanistic approach, the objects first decomposed into their consisting parts and then each part studied in separate. The outcome of this kind of thinking (also called "machine thinking"), was ignoring or less importance of the relations among the parts and in consequence ignoring the whole consisted of the related parts and the discipline ruling the object as a whole. This approach to the complex objects appeared in the 20th century due to the advancements in science and technology had many difficulties.

Development of system and systems thinking concepts resulted to the replacement of the more complete and hard – soft concept of "system" instead of the elementary and hard concept of "machine". Machine concept approximately is like the mid level systems and linear unidirectional path concept mentioned above. But the concept of systems thinking finally is what mentioned about the high level systems and circular progressive path. In other words, the "hard machine-like objects" replaced with the "hard–soft system-like objects" concept. If in the past "machine" was at the center of attentions only in its hard concept, but now "system" is in its hard and soft concept.

The new engineering approach in all fields is the result of systems approach instead of mechanical approach in the case of objects, related events and orders and rules. System and systems thinking concepts are highly used in engineering. Also, "system engineering" can be discussed from different views. From first view, as engineering of system, is as a branch of engineering with a set of engineering techniques related with systems. From second view, as engineering with system concept, indeed is the base of the new technology.

Also "systems re-engineering" is revision the structure and its order, behavior and its rule and the structural order protection and behavioral rule control mechanism and using the new scientific and technological ideas in systems. Systems re-engineering mostly introduced in the case of work and business information systems and the necessity of revisions in it according to the new advancements in the information technology. But systems re-engineering, according to its dimensions, can be discussed in the case of all the hard or soft systems.

Systems Thinking Concept

In the previous chapters, we introduced systems thinking as the combination of the two old general methods. In this book, we named these two methods as "systemic or order-based approach" and "systematic or rule-based approach". These two methods are based on the two old views as reductionism and holism. In the first method or "systemic approach" that includes "analyzing the whole into parts", "structure" and its order and in the second method or "systematic approach" that includes "synthesizing the parts into whole", "behavior" and its rule are at the center of attentions.

According to the above, in "systems thinking", identifying the parts and whole are simultaneously important. Of course, in systems thinking, the wholeness of the related parts and the discipline of the whole are emphasized. In other words, in systems thinking structure and structural order, behavior and behavioral rule and the protection and control mechanism for protecting the structural order and controlling the behavioral rule are important. For human kind, at first only the structures implying matter have been at the center of attentions and energy and information have found their roles later in the history. It can be said that in these two methods and in combination of them, first "matter", then "matter and energy" and finally "matter, energy and information" have found their roles.

As we saw, when we encounter with an event in daily living, we usually look for its cause. By finding the cause, we find that the cause is also an event due to another cause. This backward seeing and finding the previous causes if limited to one or two, may be useful, but in the complex cases, is not. Indeed, most of the daily events are an interwoven chain of different causes and effects.

Systems thinking include methods and tools that can be used for better understanding of the complex cases of problems and difficulties. Systems thinking as a "thinking pattern" imply some type of "world viewing" including seeing the things in the world around as "wholes related to each other". In other words, instead of seeing only the parts from structural view, the whole of the related parts and instead of linear and unidirectional paths, a network of circular and progressive paths of cause and effect from behavioral view is at the center of attentions.

Based on systems thinking, supposing any complex thing as a system helps to identify it by using the concepts explained in the previous chapters. For doing this, systems thinking tools can be used. In systems thinking there is a simple and natural framework for this. This framework can be used in dealing with the changes leading to transition from one step or state to another step or state. Indeed, this is the case in every day life. Hence, systems thinking can be used in all of these cases.

Systems thinking as a "methodology" for recognition of complex objects include a process with methods and tools for explanation of the systems aspects in the objects of interest. These are static structural, dynamic behavioral and disciplinal protection and control aspects in the structures and behaviors. For studying and defining the complex objects as a "system" in general, systems thinking use these methods and tools to describe the objects from the above mentioned aspects. By using the systems thinking methods and tools, we can describe the past, explain the current and forecast the future states of the objects. Theses are done by the use of general and typical properties and patterns in the structure, behavior and order and rule in systems.

It is well known that the old analytic method has many difficulties and leads to the ignorance of most of the properties of the studying objects. But the new synthetic method leads to more real and better understanding of the objects. In systems thinking, parts are important and are studied only in relation with the whole. In a study, some of the properties of the parts may or may not affect the whole as system and from this view, may or may not be important.

In systems thinking always there is a tendency from "analysis" to "synthesis". According to this, in systems thinking instead of "multiplicity" by "reductionism" view, there is a type of "unity" by "holism" view. In this tendency, "objectively due to the unity", "order and the necessity of protecting it" and "rule and the necessity of controlling it" in the whole consisted of the related parts are at the center of attentions. It must be said that being persons in some place, does not mean that there is a village or city there. Rather, if there is a village or city in some place, then there are persons there. Based on this, also in systems thinking it is believed that real cognition of some thing does not imply to emphasis on its parts, but implies understanding the relations among the parts ant its wholeness.

Of course, in systems thinking "analysis" is not ignored. Rather, in this type of thinking, "synthesis" has priority on "analysis" and in consequence, "behavior" has priority on "structure". This means that instead of thinking on "agents" (structure), it is better to think on "operations" (behavior). In other words, instead of preparing a list of agents related with some results or affecting the results, it is better to find the cause and understand how the behavior is done. This subject points to the shift from "machine" to "system" concept that has led to different perception of the existing reality. This perception results from the reality that in the machine thinking the orders and rules in the world were considered to be "constant", but in systems thinking are considered to be "variable". This is why in systems thinking, both analysis and synthesis and especially synthesis are needed.

Indeed, man in dealing with the objects has reached to "synthesis" from "analysis". As we saw, "analysis" mostly tends to "structure" and "synthesis" to "behavior". This fact is due to that the first systems man examined were systems with structural (constancy and silence) aspects. Later and with advancements in science, systems with behavioral (variation and motion) aspects also have attracted the human kind attentions.

Systems thinking can be viewed as "systems analysis and synthesis" or "top-down analysis and bottom-up synthesis" with the priority of "bottom-up synthesis" for making the whole and reaching the goals. In this analysis and synthesis, structure, behavior and discipline and the effects of the change in each of these on the others and in "stability" and "goals" of the system are discussed. In systems analysis and synthesis also the processes and the flow of input, function and output and its role in achieving the goals are discussed. If "analysis" of a whole into parts reveals "how" the whole behaves, then the "synthesis" of parts to a whole reveals "why" the whole behaves so.

In analysis, tries are made to reveal the structure and structural order of the parts in the whole. Then the "sum" of the parts studied as the whole. If whole is greater than the sum of its parts, then analysis does not lead to a true result. Because, by breaking the whole into parts, the properties of the whole mentioned above are ignored. Indeed, with analysis, the wholeness that the behavior is derived from may be ignored. In consequence, what goes in the reality can not be described truly. In systems thinking, "parts" in the form of "whole" are studied in the bed of "environment" of the whole. In other words, in addition to the above mentioned cases, the environmental factors are also important and can not be ignored.

In systems thinking tries are made to find the relations among the objects, events and orders and rules. The cases, at the time of observation or data gathering, apparently may be identified un-describable or un-related. Also it is tried to reveal the role of the bed or environment. With the old method, the past of the systems may be irrelevant and also the future of systems may not be forecasted. But by the new method, as we told, it is tried to reveal the past, describe the present and forecast the future of the systems in the bed of the environment.

In systems thinking, in studying the objects in an environment, interactions of the parts with each other and the whole with the environment are studied. The aim of this study is to reveal how the interactions can lead to total emergent properties in the whole that can not be found in the parts individually. Because, many evidences show that developing or improving the parts of a whole individually, essentially does not lead to the development or improvement of the whole consisted of the same parts.

For developing or improving the whole as a system, attentions must be paid to the "compatibility" of the consisting parts. There are many cases that using the best parts for making a whole, always has not led to a better whole. Systems thinking methodology includes the three steps of "study", "design" and "implementation" of the systems.

The purpose of "study" is to find out the structural properties, behavioral patterns and disciplinal mechanisms or the structural order and behavioral rule in the old existing or new creating systems. The study of systems is done in "system level" and "elements level". In system level study, structure, behavior and discipline of the whole system and the system goals are studied. In elements level study, study is done in the case of the elements and the role of elements in the whole system.

The purpose of "design" is to create a logical form of the system, before creating it physically. From this view, "design" includes finding and defining a desired structure, behavior and discipline in the form of a system. Design is done by using the specific tools that are available for this case. By using the design methods and tools, a complete description of the desired system in the form of text, picture, diagram, software or a "model" in general can be prepared and examined before implementing and using the system. This leads to finding and removing the errors and failures of the system before implementing and using it and in consequence, savings in the costs.

"Implementation" is a process in that the prepared plan for modification of an old existing or a new creating system is executed in practice. Implementation indeed actualizes the plan to the real system. Moving from plan or definition of a system to its implementation is a kind of "realization" of the system.

In systems thinking, in addition to the "whole", "relation" among the constituent parts and their "role" in the system and also the interactions of the whole as the "interior" with the environment as the "exterior" or the "big picture" of system including "interior and exterior", are studied. In the battle between "silence" and "motion" in the ideas of ancient philosophers, systems thinking instead of "constancy", emphasizes on "variation". Systems thinking instead of static view in the form of structure mostly tend to the dynamic view in the form of behavior. In other words, in systems thinking, structure is important because structure supports behavior. This is why in systems thinking objects without behavioral aspects, are not interested.

In different observations or studies, by finding the common properties and patterns among the apparently unrelated objects and events, specific system properties and patterns can be achieved. In finding a certain system property and pattern, by observing or studying the objects and events of interest and identifying the relations among and order and rule, first we are faced with identification of structure, behavior and discipline and finally with the definition of the related system.

In systems thinking, like the above mentioned example about the man and village or city, while considering the individual parts, we must look for a general property and pattern (village or city) that includes all the parts we want to consider. This must include certain properties from structural and certain patterns from behavioral view. It must be noted that from systems thinking view, responsibility of the behavior in any object, is with the object itself. In other words, based on systems thinking, instead of seeking the cause of the behavior of a system outside of it, first the inside of the system and then the external factors must be studied.

Systems Thinking Methods and Tools
One of first tools in systems thinking as a methodology for dealing with complex objects and related events is daily language. In systems thinking, besides using other methods and tools, ordinary daily language is also used. Daily language sentences include noun, verb, adjective (for noun as subject or object), and adverb (for verb and its space and time). By using the language, we can explain the objects (as noun in the sentence), related events (as verb in the sentence) and order and rule in the objects and events (as adjective or adverb in the sentence).

As we see, the roots of systems thinking are in the nature of the human thinking and in the form of natural language and language elements. Indeed, one of the most accessible tools for describing the objects of interest as systems in the form of system concepts is the ordinary natural language. By using the language, with no restrictions that other tools usually have, all the aspects can be explained together.

Also, abstraction, modeling and simulation are other useable tools in systems thinking. By using these tools, the object of interest can be described logically in the form of text, diagram or physically in the form of maquette, statue and other. This leads to finding and removing the difficulties of the system before creation or implementation of the system and consequently saving in the costs.

It must be added that the first step in thinking has been the use of picture and then abstracting it to alphabetic letters, words, sentences and finally language in general. Nevertheless, pictorial imaging is also an important tool in any kind of thinking. Also in systems thinking, picture and diagram are highly used.

From the very past, the hierarchy of these tools for thinking have been the use of brain without using any external memory, creating picture on the wall or on other material as the first use of external memory and finally the first pictorial languages. Then the symbolic languages and the oral or written material in the languages, picture with text or description, graph or diagram as a combination of picture and mathematics and finally modeling and simulation have attracted the attentions of human kind.

In systems with structural aspects, for displaying the system elements and relations among, "structural diagram" can be used. Traditional maps in the case of foundation, skeleton and mechanical or electrical works in buildings and in other similar cases are structural diagrams. In systems with structural and behavioral aspects, besides the use of structural diagrams for displaying the structure, also the "behavioral diagram" can be used.

Systems with structural, behavioral and disciplinal aspects and with mechanisms for protecting the structural order and controlling the behavioral rule usually are complicated. Nevertheless, in the case of these systems, also for displaying the discipline of the system, besides the use of structural and behavioral diagrams, "protection and control diagrams" and for displaying the whole system, a combination of structural, behavioral and disciplinal diagrams can be used. System researchers usually use a combination of the tools to describe their system of interest.

It must be added that for drawing all kinds of diagrams in all scientific fields, usually there are specific software tools. These software tools, in addition to presenting the required symbols in the diagrams, control the correctness of drawings. For example, if an invalid relation is created between two symbols in the drawing, the tool discovers the incorrect case. For drawing all kinds of system diagrams, also there are different software tools.

What mentioned above about explaining or displaying the systems, are about the systems that are clear enough and researcher wants to explain or display it for others. But in the case of other systems the situation may be completely different and even the researcher not know more about it. For explaining the system of interest the researcher is forced to use different tools.

Constituent elements of a system and the function of its individual elements usually may be simple and clear. Difficulty arises when there are many elements acting simultaneously. In this case, the "complexity" of the system arises from the multiplicity and diversity of the elements, relations among and the simultaneously functioning of elements and openness of the system. In other words, closed systems with no interactions with the environment, are simpler than open systems. In closed systems all the relations and interactions (if any) are limited to the inside of the system. But in the open systems both the inside and outside of the system as its environment are important. Some times, also determining the boundary of the environment of open systems is a hard work.

Researchers, study the systems according to their special situation that is called "system prospective". This is an important point and it must be noted when using the systems thinking methods and tools in approaching the objects as systems. Indeed, "system prospective" is a view in cognition of systems that is formed based on the considerations of and the available methods and tools for the researcher.

According to the above, some of the most basic useable methods and tools for describing systems are as follows:
- System Description in Ordinary Language
- Abstraction, Modeling and Simulation

Also diagrams as:
- Structural and Static Aspects Diagrams
- Behavioral and Dynamic Aspects Diagrams
- Protection and Control and Disciplinary Aspects Diagrams
- Complete System Diagrams

System Description in Ordinary Language

Language is as the "tool of tools" in any thinking activity, including systems thinking. Language includes rules for making sentences by words and assigning meaning to the sentences. In a sentence in ordinary language, none of the words in alone can carry the whole concept of the sentence. Indeed, it is the whole sentence and especially the "verb" of the sentence that carries a certain concept. This is true also in the case of a clause, paragraph, section and chapter of an article or a book. Systems thinking also confirm this view. If a sentence without a verb has not a distinct meaning, in systems thinking also the behavioral aspect of the objects (equivalent to the verb in ordinary sentence) contains the core meaning.

One of the practical methods of using the ordinary language for describing systems is so-called "brain storming" method. This is done by thinking about the subject or object and event of interest and writing any thing coming to mind about it in the form of ordinary sentences even without any organization. This writing, after initial preparation can be reviewed again and again in different times and the new things coming to mind, added to it. The first step in this work is to be certain that approximately "all the things" known about the subject or object and event of interest, are written. This writing is the base of the work and can be titled as the raw "system statement".

To continue this, it is better to draw three tables with a number of rows and columns. Put the title of first table as "objects and its structural orders", second table as "events and its behavioral rules" and third table as "related objects and events". Then extract all the nouns or noun implied words from the "system statement" and put a side all the cases that are not related with the subject or object of interest. Now, among the extracted and refined nouns or noun implying words, determine common and proper nouns and each proper noun under a common noun.

Then write the common nouns in bold face and the related proper nouns in ordinary face beneath the related common noun in the cells of the first column of the first table. This is done for all the common nouns and related proper nouns until all the extracted nouns are written in the first column of the first table. Then find the adjectives of the nouns in the system statement and write it in the cells after the related noun in the rows of this table. Now the rows of the first table contain the nouns and adjectives related with the nouns.

The same work is done in the case of compound verbs (like common nouns) and the related simple verbs beneath it (like the proper nouns related to each common noun) and adverbs (like adjectives of nouns) related with it. In other words, like nouns and adjectives in the first table, verbs and adverbs are written in the second table. After doing this, all the compound verbs (in bold face) and the related simple words beneath each related compound verb (in ordinary face) are written in the first column of the second table. Like before, the related adverbs placed in the cells after the verbs in the rows of the table. Now the rows of the second table contain the verbs and adverbs related with the verbs.

In the first table, in the case of nouns determine if the noun is subject or object. Display this by writing "S" (for subjects) or "O" (for objects) in the left-down corner of the related cell. Also in the second table, in the case of verbs determine the time of the verb. Like before, display this by writing "Ps" (for past), "Pr" (for present) and "Fu" (for future) in the left-down corner of the related cell.

For completing the work, combine the first and second tables by intersecting and write the result in the third table. In other words, in the third table, assign the rows to nouns (written in the first column of the first table) and columns to verbs (written in the first column of the second table). Then copy all the nouns in the first column of the first table to the first column and all the verbs in the first column of the second table to the first row of the third table.

Now by cross referencing the nouns and verbs in the rows and columns of the third table, determine if a noun (as subject or object) is related with a verb and put an "X" in the intersection of the related row and column. Now it can be said that from the written "system statement", a raw list of objects (nouns as subject or object), related events (verbs) and its orders and rules (adjectives related with nouns as structural orders and adverbs related with verbs as behavioral rules) is extracted.

In the extracted cases, approximately each common noun can be viewed as a "group of objects" or system. Also, approximately each proper noun under a common noun can be viewed as an "object" or an element of a system. Like this, approximately each compound verb can be viewed as a "function" in the form of a process in a system and approximately each simple verb under a compound verb, can be viewed as one aspect of the behavior under the function of the related system.

To find the cause and effect relations or processes, in the third table, especially the subjects (an aspect of cause), verbs (as event due to the cause) and objects (an aspect of effect) can be helpful. Also, approximately it can be said that the "history of the system behavior" is extractable from the past tense verbs, "current behavior of the system" from the present tense verbs and "forecasting the future behavior of the system" from the future tense verbs.

Studying common structural properties and behavioral patterns among the groups of objects and finding a common subject in structure or behavior covering all the objects in the group, is another method in approaching to a system. In this case, indeed, the common subject will include a system that approximately can contain all the objects in the group. For doing this, the system statement or the three tables discussed above can be helpful.

Now the structure and structural order, behavior and behavioral rule and the protection and control mechanism for protecting the structural order and controlling the behavioral rule or the discipline of the system also can be explained. As we see, if a certain object or subject of interest is explained in a "certain" way, then we can identify the related "system". As we saw in the above, this is done by writing the system statement mentioned above, then extracting the "system concepts" and finally "combining" them in order to achieve to a "whole" as a system.

It must be added that in principal, the language, sentences, words used in a sentence and especially "noun", "verb" and "adjective and adverb", naturally and mostly agree with what is told about the "object", "event" and their "order and rule". In other words, in natural language, "object" is stated in the form of "noun" and "event" in the form of "verb". Also, "order" as "property" or "how the noun is" is stated in the form of "adjective" and "rule" as "method" or "how the verb done" in the form of "adverb".

By applying the structural aspects in the system concept, processed "system statement" can have a certain structure and used as system explanation. The written system statement in a well formed state can have the following structure:
- Starting material
- Body of the text
- Ending material

Starting material can include the followings:
- Title: Title must be short and possibly reflect the content.
- Writer or writers
- Keywords reflecting the content: "Keyword" is a word that reflects content in a way. Keywords can be grouped in two general groups as "main" and "subsidiary". "Main keywords" reflect the content "explicitly" and "subsidiary keywords" reflect it "implicitly". Keywords must be written in the order of their importance in the content, to notify the reader about the importance of them in the content. Hence, writing at least two keywords including the main (reflecting the main subject) and subsidiary (reflecting the subsidiary subject), can be helpful.
- Abstract: "Abstract" in general, includes a compressed statement of the whole content by the style used in the content, regardless of the size or structure of the whole content. Abstract must be written without any interpretation or criticizing the original content and with the least number of sentences must include the important key points. The goal of abstract is only informing about the original content in order to help the reader in a short time to decide about the reading of the whole content.
- Summary: "Summary" in general, includes a shortened content of the whole content by the style used in the content and regarding the structure of it from the original sections view. Summary must be written without any interpretation or criticizing the original content and it is better that each section of it to be the abstract of the related section in the original content. The goal of summary indeed is presenting a shortened content of the whole content for the people who do not have enough time to read the longer contents.
- Table of contents

The body of the text can be organized in different sections (including structure, behavior and discipline of the system of interest) and each section in few paragraphs. It is better that each paragraph to include a certain discussion about the subject. Also, It is better that the sections of the whole text to be organized in two levels including a few main sections and each main section in a few sub-sections. Here, like the main title, titles of the main or sub-sections must be short.

Ending material can include the followings:
- Conclusions
- Further studies and researches for more information about the system of interest
- Glossary of the technical words used in the text, in alphabetical order
- Appendices, including the material not appeared in the text, but useful for better understandings.
- References, including author, date of publishing, title of the reference, publisher's name and publishing place and if necessary, the referred pages. Direct referring to the references are made inside the quotation marks ("") and the number of the reference in the references list inside the () or []. In the case of internet sources, date of publishing is the date of access, and publishing place is the related internet address.

Abstraction, Modeling and Simulation

"Abstraction" in general, is one the methods used in systems thinking. Here, abstraction is considering the limited general, instead of numerous special aspects in the object of interest. By doing this, a complex reality is displayed simply. In other words, in abstraction we are faced with the basic and principal aspects of the things.

In the abstraction process, keeping out of the special aspects, leads to high rate of abstraction. Moving from special aspects towards the general aspects or having more abstraction is considered as moving towards a kind of "generalization". In contrast, moving from general aspects towards the special aspects or having less abstraction is considered as moving towards a kind of "specialization".

"Model" is an abstraction of an object regardless of its details. In other words, model is a summarized and simplified form of an object and is created in order to better understand it. In a model, emphasizes are on those aspects of the object that are important in studying it. Model is made in order to study the object by attention on those aspects. Model can be in a logical form like mind imagination, picture, diagram, mathematical relations, plan and other or in a physical form like maquette, statue, tool or device and other. The process of creating the model is called "modeling".

Also "object model" is some thing that indicates the object in the same form as it is. Object model can indicate the constituent elements of a system or the elements that are used by the system. Preparation of object model is one of the first activities in identifying the object or related system. From systems view, model of a system is a simplified form of the system and indicates the most important aspects of it that can be studied. The aspects include structure, behavior and discipline of the system or a combination of these three that are important from the researcher's view.

Hence, "System model" is the "abstraction of important aspects" of the system of interest in order to understand it according to its researcher's goals and priorities. Important aspects are those that are important in the field of study or from the researcher's view. In the case of a system, the important aspects may vary in different studies and times or from different researcher's views. In the model of a system, the priority in abstraction may be the structure and structural order, behavior and behavioral rule or the protection and control mechanisms for protecting the structural order and controlling the behavioral rule or a combination of these.

The value or importance of models, regardless of the apparent shape of model, is that it shows the object of interest from one or more sides. In model, instead of facing with the complex world of reality, we are faced with a simpler world limited to the parts or components that are important from the view of research. In order to explain a system, besides describing it in ordinary language, modeling methods also can be used in order to make a model of the system.

One of the methods for studying an unknown thing is to find its similarities with another known thing. Of course in any case, analogy or resemblance usually is made only on some properties. According to the good or not good cases of analogy or resemblance, this may help the researcher and lead to better understanding of the thing or in contrast mislead the researcher. If some thing is not more similar with the experiences, then a distinct case similar with it may not be found. In this case, also the model may be helpful in a limited way. Such a model also may be very simple. But, using a model even a simple one may be very hard in practice.

The accuracy of the model of complex systems, because of the many related varying factors, usually is limited. An abstract model may be good form observational views, but meaningless in its nature. Of course models may contain some unwanted meaningless aspects. From researcher's view, some of the events like the growth of plants are slow and some others like the car accidental collisions are fast. In these cases, the model must speed up or slow down the related event in order to study it.

As we saw, model of a system can be created from different views including structure, behavior and discipline or a combination of these. According to the case, "structure model", "behavior model", "discipline\ protection and control model" or a set of all as "system model" may be created. According to what told about the model in the above, models can be classified in different classes.

A type of classification may be as follows:
- Structure and structural order models, like picture, structural diagram or maquette with fixed parts for displaying the structure and structural order of the system
- Behavior and behavioral rule models, like behavioral diagram, maquette with moving parts for displaying behavior and behavioral rule of the system by displaying motion and change in the diagram or in the moving parts in the maquette. In this case, the structure may be summarized in its least form to support the behavior.
- Disciplinal or protection and control models for protecting the structural order and controlling the behavioral rule, like the combination of structural diagrams reflecting the structural order and behavioral diagrams reflecting the behavioral rule with the flow of "information" besides "matter" and "energy" for displaying the disciplinal aspects in the system of interest and emphasizing on how the structural order is protected and behavioral rule is controlled.
- Combination of the above mentioned models as a complete model for a system

Following shows the simplicity or complexity of systems modeling:
- Simple modeling: Fixed structure systems as static systems
- Some how complex modeling: Systems with discrete or continuous structure and behavior as dynamic systems
- Complex modeling: Systems with discrete or continuous structure and behavior and discipline including structural order protection and behavioral rule control as organic systems

Models according to the apparent form of the model can be classified in three main classes as "physical model", "logical model" or "physical – logical model". Model can be in a physical form as maquette, tool or device and other or in a logical form as picture, diagram, mathematical relation and other.

Physical models are "matter based" and "sensible" or "hardware" as small samples of the real objects. It is clear that using a model of a car for accidental collisions in order to study the effects of collision is least cost and safer than using a real car in a real collision. Unlike physical models, logical models are "conceptual" or "non physical" and are made in "understandable" or "software" form.

Today, computer based logical modeling is the most common method in modeling. Logical modeling compared with physical modeling, has more flexibility and low cost. This is why in the case of most systems, instead of physical modeling, computer based logical modeling is used. Computers by running more complex programs and executing many instructions in few seconds can deliver the results in shorter times. In consequence, processes are modeled in the same way that we want or imagine.

Most of the logical models can be abstracted in the form of mathematical relations and used to forecast the future behavior of the systems. In most cases, a mathematical relation that is valid in a specified range of variations of the related variables may lose its validity in a wider range of variations of the same variables. Different parts of systems can be related with each other and in different situations, the behavior of systems and its error and failure can be modeled by mathematical relations. Among the many types of models, mathematical models are made with the most possible abstraction of the real world.

Logical models with behavioral or disciplinal aspects in systems also may include a set of exact rules and instructions in order to determine the steps that must be done. In some cases that performing of real steps are very difficult, even a simple set of rules and instructions may lead to better results.

In systems thinking, while using a level of abstraction in modeling, the patterns of models are taken from the world around. Indeed, logical models are a display of the object of interest with a least and key structure and behavior that can not be disregarded in it. After making the model, it must be compared with the real world and the similarities and differences between the model and real world must be identified. Then the desired changes in the model must be determined and finally applied in the structure and behavior of the model.

Since the study of complex systems needs some kind of modeling, understanding the basic concepts of modeling can be helpful in systems studies and modeling. As we saw, there are different methods for systems modeling. Regardless of the method used for modeling, model must be alike and similar with the system that the model is making for it. Studying the system of interest and determining the best method of modeling for it, is one of the first activities that the researcher of a system may be faced with. Best model for a system is different in the different situations. If the general aspects of the system of interest are not well known or extracting mathematical relations is difficult, then physical modeling may be better.

It must be said that the information about a system in its researcher's mind, usually is more than the information documented by the researcher about the same system. Now if a model restricts the information about the system only to the quantitative measurable information, then this information will be lesser than the documented information about the same system. In other words, while subjective models contain more data than objective and documented models and also documented models than quantitative models, in practice there is no way other than to document or make quantitative models about the systems.

In quantitative modeling, no details are needed. For example, in the case of properties of an object, instead of qualitative properties, only general quantitative properties are enough. Quantitative modeling usually meets the quantitative purposes and if possible reflects some qualitative aspects.

Human mind is good in making new patterns but weak in breaking or putting aside the old ones. Here, innovation as "the ability of breaking or putting aside the old traditional patterns and presenting the new ones" better and simpler than the old ones can take place. In the first step, innovation is determining the cases that cause to neglect the new cases other than the old ones we are faced with. Some times, these cases may act as self-wanted and not the real restrictions. For innovation, the ability of breaking the restrictions and studying the results is necessary. Innovation can be improved by experience and pattern making.

The process of model making is "iterative and progressive". This means that, first the "structure" of the model is made and then the "behavior" of it is tested and compared with the real or expected behavior. The result of the test and comparison is used again as feed-back for improving the created structure. This loop is repeated until the difference between the modeled and real behavior is minimum or acceptable. This implies some kind of returning back, reviewing the previous steps and modifying it to achieve the expected results.

It must be said that some times the difference between the real and modeled behavior may be great. This may be the result of error or failure in the modeled structure and behavior or in contrast a sign of an old or a new aspect of behavior not considered or thought about before. If the difference between the modeled and real behavior is not resulted from the error or failure in the modeled structure and behavior, then it may be related with a new or unknown aspect of the behavior that must be studied.

Model also is as a device to transfer the knowledge and share it with others. Model must be understandable by the model presenters and especially by the audiences in discussions and negotiations. Important thing in understanding the model is that all the recipients of the concepts presented by the model must get the same understanding of the discussing subject.

Model must play its role that is simplifying the complexities and transferring the concepts to the others. Even the complex objects must be simplified and explained in an understandable form. If a model presents full percent of a real object, usually is not a good model. One reason is that in most cases, even in the case of a simple object, all the aspects are not important for the researcher of the object, because usually the discussion about it is limited. Other reason is that, in the case of complex objects, human mind can not deal with the whole object in one step. In modeling a complex object, it may be broken into smaller parts and then by making and combining the models of parts, the model of the whole object to be created. In any case, model must deal with the important aspects in the studying object.

One of the structural properties of a model is having a "theme". This property is like the theme in a story that the writer organizes the story around it. Theme in a model causes to determine a "boundary" for it and identify the cases that belong to the inside or outside of this boundary. Of course, model must deal with the cases belong to the inside of this boundary and ignore the cases in the outside except the cases that are as the environment of the system.

For making a model, simultaneous attention on "whole" and "parts" is required. According to this general method, first step for making a model is abstracting the real object and determining the theme and identifying its domain. By determining the "main parts", also belonging or not belonging of the parts to the domain of the model must be determined. Abstracting the real object to its main parts indeed makes its "big picture" from our view. In this picture, all the main parts of the object and relations among are displayed in the bed of its environment. Unlike this, determining the main parts and in consequence the whole model, leads to identifying of the domain of model. Also, determining the domain in a model, leads to identification of more elements in it. By identifying the constituent elements of a domain, it will be seen that there are also certain relations among them.

In the example of filling a cup with coffee from a kettle, by keeping the state of kettle by hand and pouring the coffee to the cup, the level of the coffee in the cup continuously controlled by eyes in order to prevent the overflow of coffee from the cup. In this case, when there is a certain level of coffee in the cup, pouring of coffee from the kettle is stopped. Experiences show that most of the human activities are in the form of a model like this. This model is an abstraction of what occurs in the real world and in it "a structure wants to achieve a goal by a behavior".

"Structure" is a base that other things are achieved by it. "Goal", like filling the cup with coffee, usually is a final state that must be achieved. "Behavior" determines how to pass the mid states in order to achieve the goal as final state. In the above example, the goal of behavior is pouring coffee to the cup and finally drinking it. Mind and eyes continuously monitor the whole process including keeping the kettle in hand and controlling the level of coffee in the cup. This model is also true in systems.

All the processes related with the specific work activities contain the main components including structure, behavior and discipline (including protecting structural order and controlling behavioral rule) in order to achieve the goals. In most cases "structure" is supposed to be constant and the other two aspects or "behavior" and "discipline" supposed to be variable. For example the goal of any business usually is to earn profit. For this, the structure of business by behaving in a way, wants to maximize the profit. This is why, in system models, the required structure, behavior and discipline to achieve the goals, are important.

Based on this, "simulation" is a process for modeling or explaining the structure, behavior and discipline of a system in a logical or virtual form. Systems thinking, instead of using traditional or ordinary methods, widely use "computer simulation" methods. Simulation is done by the use of a set of structural constants and variables and functions of behavioral patterns and goal.

This set represents the structure, behavior, discipline and goal of the system in a logical form, and by it, a clear understanding about the system becomes available. In other words, simulation of a system is a process in that, while examining the function or behavior of the system for achieving the goals, also other unknown aspects of the system can be studied.

Simulation can include discrete or continuous behavior. In discrete behavior, the system of interest, in responding to certain internal or external events, goes from one state to the other state and this occurs in certain moments of time. But continuous behavior is evolutionary and can not be breakdown easily into separated states or steps.

One of the effective methods of modeling for system studies is "visual modeling". Visual modeling is the use of symbols and pictures in making models and in consequence, easy understanding of the system of interest by others. Each pictorial tool is a way for extracting the mental ideas and thoughts and sharing others in understanding the system of interest.

A set of signs and symbols in a certain case, usually can be proposed in the form of a system itself. In this case, the "relation" among the signs and symbols, transfers the knowledge about the objects to the others. In this information interchange, there are three sides of "object", "sign" and "receiver" and the whole set of it can be viewed as a system. It can be said that a "receiver" is informed about a certain "object" by the use of certain "system of signs". Basic discussions about symbols, includes "analysis and synthesis" or cases related with the form or apparent and content or concept of the symbols.

Diagrams
The pictorial or graphical methods discussed in this book, are among the visual modeling methods in the case of systems. A full set of these diagrams are also available in another volume of this book with sub title of "Work and Teaching Book". Readers after studying the current book can refer to this set of teaching material.

For displaying the structure and static structural aspects of the systems in a simple and general way, among the many "structural diagrams", "picture diagram" or "block diagram" or a combination of these two can be used. Also for displaying the behavior and dynamic behavioral aspects, among the many "behavioral diagrams", diagrams with the general title of "flow diagrams" can be used.

Of course any behavior and behavioral rule diagram can be based on a structure and structural order diagram mentioned above. Flow diagrams include "flow chart", "activity diagram", "states chart", "Finite State Machine\ Model = FSM", "cause and effect diagram", "stock and flow diagram" and especially the "process flow diagram" including "Input – Process\ Function – Output = IPO\ IFO".

For displaying the disciplinal aspects in systems, including the mechanism of protection and control for protecting the structural order and controlling the behavioral rule, a combination of the above mentioned diagrams as "protection and control diagram" can be used. For this purpose, closed loops of information feedback are used in the diagrams.

Also for displaying the whole system, a combination of structural, behavioral and disciplinal diagrams and especially with the centrality of behavior in systems, "process flow diagrams" can be used. According to the conditions or requirements, structural, behavioral and disciplinal diagrams in systems can be presented in different levels. In this case, diagrams can be presented in different layers including the structure as the basement layer and behavior (on structure) and discipline (on structure and behavior) layers.

System researchers may use writing and a combination of the above mentioned diagrams for explaining their system of interest. In explaining a system in writing form like an article, in different situations and for different purposes, usually "detail", "summary" or "abstract" can be presented. Also systems diagrams can be presented in three different levels as above.

As we saw, the abstract of an article, without regarding the structure of the whole article, only includes the key points in one or few paragraphs. While summary includes the abstracts of sections under related titles and with the same structure of the whole article. Detail of the article includes the whole writing with its whole structure. Like this, also the systems diagrams may be presented in different levels containing "sub-sections" (detail), "main-sections" (summary) and "key-sections" (abstract). This is like presenting the human body in the whole form, main organs or only the skeleton of the body.

Systems Thinking and Soft\ Hard Systems

As we saw, system and systems thinking as the theory and practice for analysis and synthesize of objects and their related events as systems can be used approximately in any field. System and systems thinking present a common platform for understanding how the objects due to the related events change in time. It must be added that in the past, most of the methods in most of the cases were limited to the numerical measurable aspects. By this, a vast amount of descriptive knowledge about the studying objects or related events were ignored. While, systems thinking use all the existing information in a certain framework.

In the past century, in discovering and developing the existing physical sensible world, in the framework of hard quantitative system concept, significant achievements were made. Now, also in the case of conceptual understandable world in the framework of soft qualitative system concept, significant achievements have been gained. Unlike hard systems, in soft systems the consisting elements, relations among and the system boundaries are not clear and can not be determined easily. Soft systems researchers are faced with ambiguity in parts, whole and goals and structure, behavior and discipline of the under study system.

Due to uncertainty in the case of boundary of soft systems and depending on the attitude or views on the system, the supposed boundary for a soft system may change easily. Usually, there is not a common agreement on considering a certain boundary for soft systems. If uncertainty in boundary is the case in soft systems, then "hardness" and "softness" also become relative aspects. Because, when conditions change, the former hard systems may become new soft systems.

In other words, in soft systems, the boarder between the inside as system and outside as environment, is uncertain. For this reason, in the structure, behavior and discipline of soft systems, there are many different factors that can not be identified and if identified, can not be managed easily. It can be said that, in the case of elements, relations and boundary in systems, being quantifiable or unquantifiable is some how related with being identifiable or unidentifiable. Hard systems with certain elements, relations and boundary, have more quantitative aspects and soft systems with no such properties, have more qualitative aspects.

In consequence, solutions for problems and difficulties of the soft systems are not clear and defining the goals of the system is part of the subject of study. By changing the researchers or their views about the system, the problems and difficulties also change and some times become as opposite. For example, consider a factory producing different products. The only duty of this factory in the past was to produce products and to present it in the market. But today, the same factory in the market is faced with many rivals, variety of similar products, different customers with different tastes, amount of customer satisfactions from the services during the sale and after sale and many other similar cases.

Above mentioned cases are so important in economics today that some times even least neglects may lead to huge losses and exiting from the market even for the giant active in the market. Today, in studying this kind of systems that viewed "hard" in the past, there are many new unknown factors. In consequence the boundary of the system also becomes unclear and ambiguous.

There is disagreement on supposing this idea that the hard physical and soft conceptual systems are the same in base but different in complexity. In complex soft systems like social systems, there are chains of causes and effects. In such a situation, "effect" is affected by "cause" and the "cause" in its turn, affects the "effect", both in different ways. Based on this, soft systems can not be approached as easy as hard systems.

From information and protection and control mechanisms view, the mechanisms in soft systems are not as easy and effective as hard systems. For example creating a protection and control mechanism for a hard system like an ordinary or electronic door is easy and effective without any side effects. But, we remember the huge side effects of such mechanisms in the case of social communities appeared in some countries between or after the two world wars. For example, emphasis on militarized orders to protect social structure (like in Latin America countries) and emphasis on socialized rules to control social behavior (like in socialist countries) individually or both together, are among the many in this case.

As much as the system is soft, it is more difficult to study, design and implement it. In consequence, achieving to a concrete and complete system in one step is very difficult or impossible in practice. A general true method to approach soft systems is that, the soft systems first must be studied from different points of views. Also, basic definitions and conceptual models may be used for explaining soft systems. Only in this way the other complementary efforts may be useful and effective. In practice, there is no way only to choose an evolutionary stepwise method in this case.

For systems approach in the case of soft systems, the following stepwise method may be useful:
- Determining the case or "why is so?" in the case of an object, event or the order and rule in it
- Trying to determine the structure related with the case
- Trying to determine the behavior related with the case
- Trying to determine the discipline in the case in the form of order in the related structure and rule in the related behavior
- Trying to determine the dependencies between the structure and structural order and behavior and behavioral rule
- Trying to determine the methods of protecting the order in the structure and controlling the rule in the behavior
- Determining some aspects of the behavior as the "cause" in the under study case
- Determining some aspects of the structure related with the aspects of behavior determined as the "cause"
- Making a model for a system related with the case, with emphasizing on its wholeness and structure and structural order, behavior and behavioral rule and the methods of protecting and controlling it in the whole or in a section of interest in it
- Testing the model, troubleshooting and modifying it
- Being sure that the created model represents the real behavior
- Repeating the above steps again in uncertainties or if required
- Conclusions and applying it in practice

Extra Teaching Material
for
Instructors and Students

Table of Contents

Our World
Our world is mainly consisted of "objects" and "events".

Object
Any objective or subjective thing that some how exists in "space".
Objects are the "constant" aspects of our world.
Example
A person as a living thing
A tree as a plant
A piece of stone as a solid thing

Event
Any objective or subjective thing that some how occurs in "time" in relation with the objects.
Events are the "variable" aspects of our world.
Example
Coming or going of a person
Turning to be green or yellow of a tree
Changing the place of a piece of stone

Object's Order
Objects are consisted of a set of "related parts" as a "whole" with their specific "order" besides each other.
Example
Parts of a car or a device connected to each other

Event's Rule
Events in relation with the objects have their own specific "rule".
Example
Rising or setting of sun or changing the seasons on the earth.

Evidences Confirming Above Mentioned Concepts

Based on the knowledge today, the universe itself has come to existence form a "primary condensed matter" as an "object" in the primary space and "primary great explosion" as the "event" in the primary time or "big bang".

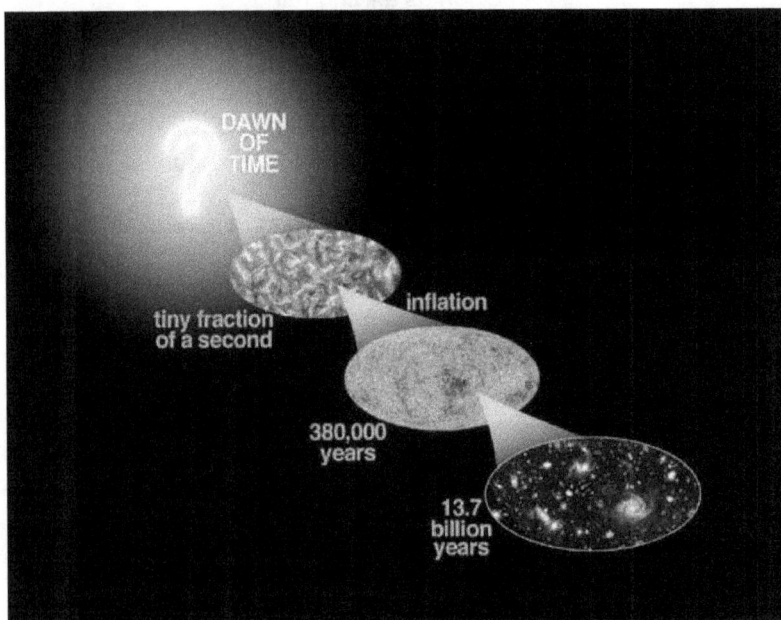

http://www.nasa.gov/centers/goddard/images/content/96118main_Mysteriesm.jpg

Big Bang, start of expansion and a picture of world today

Based on the existing scientific evidences, universe in the form of matter in it is expanding in the dimensions of space and flow of time.

In natural languages as one of the first and most common available tools for explaining the world, "noun" as one the most important language elements, points to "object" and "verb" as another important language element, points to "event".

In a sentence in language, like the objects and events in the ordinary world, nouns and verbs are not considered separately but together have a certain concept.

Like the role of events in the case of objects, what is challenging in a sentence in language, is the verb of the sentence.

Sentences containing nouns and verbs usually are important due to the verb of the sentence.
So, sentences without verbs are unable to transfer a concept.

In a sentence,
each noun can have an "adjective" that explains a certain "property or order" of the related object
and
a verb can have an "adverb" that explains a certain "method or rule" of the related event.

Adjective of a noun, some how points to the order in the noun that points to the object and adverb of a verb, some how points to the rule in the verb that points to the event.

As a result,
in natural languages, noun points to object, verb points to event, adjective points to the order governing the object and adverb points to the rule governing the event.

If in the ordinary world, objects are related with events, in natural languages nouns are related with verbs. If an object without a related event has no challenge, also noun without a related verb has no challenge.

Structure, Behavior and Discipline
in
objects and their related events

As a result of above mentioned points
Objects have their own specific "structure" in space and due to the related events, have their own specific "behavior" in time.
A specific "structural order" governs the structure of the objects.
Structural order some how "is protected".
A specific "behavioral rule" governs the behavior of the objects.
Behavioral rule some how "is controlled".
Structural order and protecting it in space and behavioral rule and controlling it in time in total, represent the "discipline" of the objects.
Objects in the form of their structure, behavior and discipline, have their own "wholeness".

Structure

Structure gives existence to the objects.
Structure is space depended and includes some aspects of "constancy" or "silence" in the dimensions of space.
Structure has a base role.

Behavior

Behavior is to "function" in order to reach to the "goal".
Behavior is time depended and includes some aspects of "variation" or "motion" in the flow of time.
Behavior is done based on the structure.
Without structure, no behavior is possible.
Structure and its order and behavior and its rule are related with the concept of "organization" and protecting the structural order and controlling the behavioral rule are related with the concept of "management", both the most important subjects in the contemporary world.

Discipline

Any structure or behavior is supposable only under a discipline.
Discipline has space and time aspects and includes some aspects of constancy or silence and variation or motion in space and time.

Structure, Behavior and discipline of the Universe

"Structure" of universe includes the matter existing in it and its "structural order" includes the arrangement governing the matter in space.

"Behavior" of universe includes expanding and its "behavioral rule" includes the method governing the expanding of the universe in all sides in time.

"Discipline" governing the universe in the form of "discipline governing the space and time" indeed are the natural laws governing the structure and behavior of the universe that science always seeks to find it.

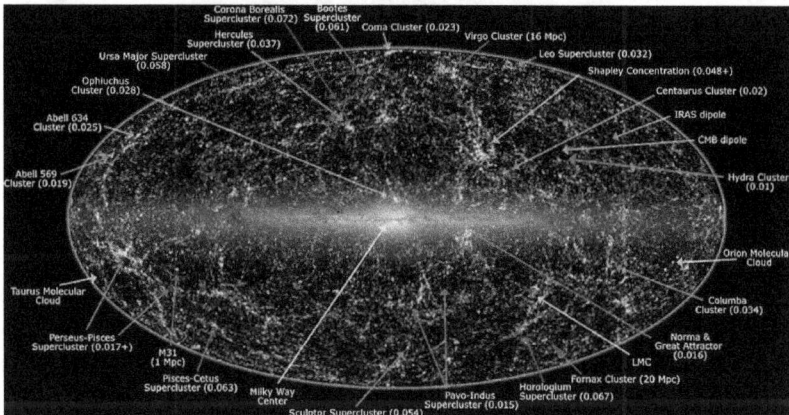

http://upload.wikimedia.org/wikipedia/commons/7/7d/2MASS_LSS_chart-NEW_Nasa.jpg

A panoramic view of the entire observable sky

Structure, Behavior and discipline in Daily Language

In daily language, "noun" reminds the "structure" and "verb" reminds the "behavior" of the object.
Also,
"adjective" and "adverb" remind the "discipline" governing the object.

http://en.wikipedia.org/wiki/File:Simetria-bilateria.svg

Any object in nature has structure with specific order or property in its form or shape.
Example: Symmetry in the body of living beings

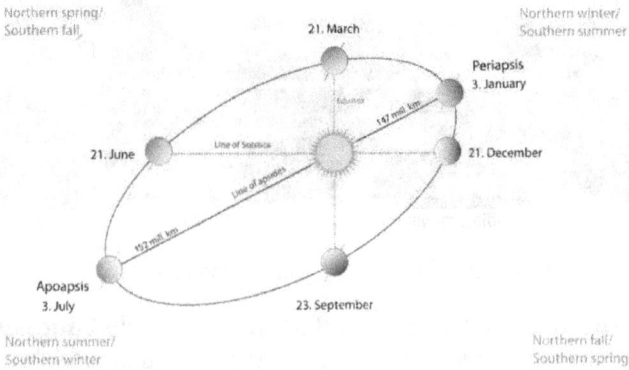

http://en.wikipedia.org/wiki/File:Seasons1.svg

Any event in nature has behavior with specific rule or method in its content or function.
Example: Occurrence of four seasons in earth

Matter, Energy and Information
(three main and basic resources)
Structure is possible by "matter" and behavior by "energy".
Matter is synonymous with mass and energy with force.
Discipline in the objects with related events, is possible in the following cases:

- From the structural order due to the matter or behavioral rule due to the energy in the form of "field" in physics or "environment" in general.
- By the aid of "information" in the objects that can use it. From this view, information is in the same level as matter and energy.

Example

According to the laws of physics, mass is the agent for structure in space, force is the agent for behavior in time and gravitational field due to the mass or force that governs the space and time, is the agent for discipline or protecting the structure and controlling the behavior of the celestial objects like the moon and its rotation around the earth and the earth and its rotation around the sun.

Information is the reflector of constancy and variation and finds its concept by the fix and change in the structure due to matter and specially behavior due to energy.

By adding information to matter and energy, more complex forms of discipline become possible in the objects that can use information.

Other role of information in the objects is its role as a descriptor of the structure due to matter and behavior due to energy.

Information is related with the usually constant structural properties and usually variable behavioral situations in the objects and so describes the case in some way.

By the aid of information, in addition to objective things, subjective things also can be explained and existed in a way.

Life of the living beings or function of the objects that can use information is not possible without the accessibility to some forms of information from the environment.

So,

matter, energy and information as the three main and basic resources, are the generator of structures, motivator of behaviors and establisher of disciplines.

Objects according to their type, for existing or to continue existing, are based on or need these three main and basic resources.

Following table displays the significant and epochal role of matter, energy and information in the living of human kind on the earth.

Fundamental Development Eras in Human's Life			
Era	**First Steps**	**Mid steps**	**Final Steps**
Matter	- Making Stony and Wooden Tools	- Discovering Metals	- Discovering New Elements - Making Artificial Elements
Energy	- Using Fire	- Using Water and Wind Energy - Making Steam Engine	- Discovering Electrical and Atomic Energy
Information	- Appearance of Language, Counting and Writing	- Appearance of Clay or Stony Tablets - Inventing Paper and Printing	- Inventing Telegraph, Telephone, Television, Satellite, Computer and Internet
Turning points of different eras in human's life. "First steps during thousands of years", "Mid steps during hundreds of years" and "Final steps during tens of years".			

Less Challenge, More Challenge
In most cases, not the objects implying the constant aspects, but the related events implying variable aspects or "objects as subjects of events", are among the most challenging cases in our world.
Naturally, in dealing with the objects implying the constancy and silence aspects, there is less challenge and in dealing with the events implying the variation and motion aspects, there is more challenge.

Some Points from the History
Along the history, man has examined first, the objects and their governing orders and second, the related events and their governing rules.
Oldness of "Geometry" respect to "Algebra" perhaps can be a reason for this, because Geometry dealt with the constant aspects or structure, and Algebra dealt with the variable aspects or behavior of the objects in the world around.
Man had been more successful in dealing with the objects than dealing with the events.

Spectrum of the objects in nature
(Solids to plants and animals)
In one end of this spectrum (solids), usually there is a matter – energy bed (some thing like field in physics or environment in nature) that encompasses the object and usually causes a constant structural order and behavioral rule in it.
Example
a piece of stone silent in its environment in the nature has its own constant structural order. Also moon by being in the field of energy or force of the gravity due to the matter or mass of the earth has its own constant structural order (as a sphere) and behavioral rule (as rotating around the earth).

In the other end of this spectrum (animals and especially human kind), discipline is possible by interacting with the environment and by the use of information as the third agent.

System

By attention to the above mentioned basic concepts
and
based on the knowledge today
"System"
is any objective or subjective thing consisted of "related parts" as a
"whole" that has its own "structure", "behavior" and "discipline" in order
to have a certain "application" or to achieve a certain "goal'.
The "wholeness" of system is greater than the sum of its constituents.

System in Different Interpretations

In some cases, the purpose of system is one of the structure, behavior or
discipline aspects.
In some other cases, a combination of these three including: structure +
behavior,
structure + discipline (structural order of discipline),
behavior + discipline (behavioral rule of discipline)
or finally
structure + behavior + discipline (structural order and behavioral rule of
discipline).

This is why
all the meanings of "system" in most dictionaries are the synonyms or a
combination of these three base concepts.

System in Elementary Interpretation

"System is a set of certain objective or subjective related elements as a whole in the form of an object in general that has a structure with certain order".

In this case, objects as systems, for their researchers are important only from the view of their certain "structure".

This kind of systems also can be called as "static system".

System in Intermediate Interpretation

Objects usually are the subject of events and due to the events, have some kind of behavior.

So, objects and related events must be considered together.

In other words,

objects with certain structure, due to the events occur for them, have their own certain "behavior".

As a complete concept respect to the previous concept, "system is a structure with certain order that due the events occur for it, has a behavior with certain rule".

This kind of systems also can be called as "dynamic system".

Egyptian Pyramids as "structure" in that thousands of pieces of stone, related with each other in a certain order, to form a pyramid as a "whole", can be viewed as a "static" or "structure" system.

Eiffel Tower as a "structure" in that thousands of pieces of metal in different sizes and classifications, related with each other in a certain order, to form the "Eiffel Tower" as a "whole", can be viewed as a "static" or "structure" system.

Wind mill as a "mechanism" (structure with certain behavior) that is one of the first man made devices to make flour from wheat by the use of energy of wind, can be viewed as a "dynamic" or "structure with behavior" system.

Clock as a "mechanism" (structure with certain behavior) that is one of the first man made machines to measure the time (yeast of any behavior) by the use of saved energy, can be viewed as a "dynamic" or "structure with behavior" system.

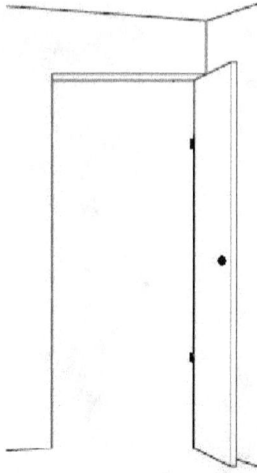

An ordinary door as a "mechanism" (structure with certain behavior), due to the leaf that can be opened or closed by the energy of the passer (event for door), can be viewed as a "dynamic" or "structure with behavior" system. The "order" and "rule" that governs the structure and behavior of the door are presented by its traditional shape (frame, leaf and hinges) and functioning (opening and closing). Frame is the basic structural part, leaf and hinges are behavioral parts and knob and lock are disciplinal parts of the door.

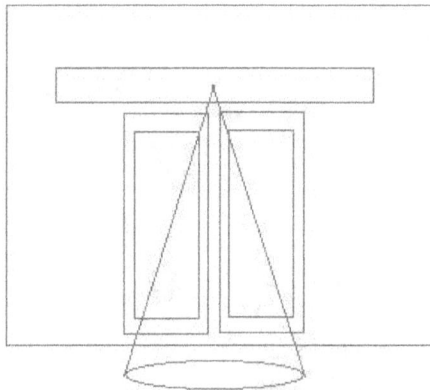

In the electric doors, behavior of the door is possible by the use of electric motor and energy and the rule governing the behavior of door, by the electronic sensor. Here, electric motor is the structural part of behavior and the electronic sensor is the structural part of discipline of the door.

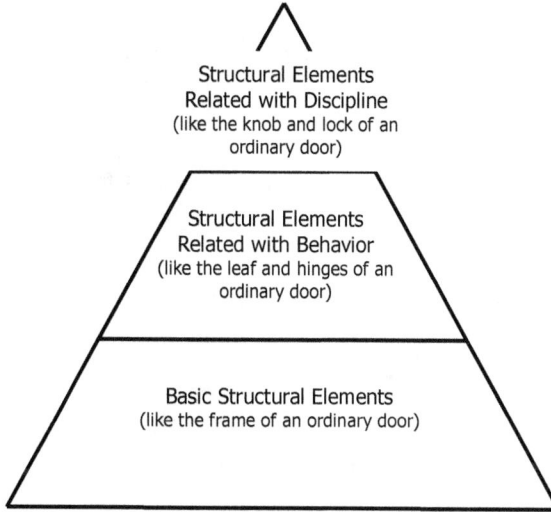

Hierarchy of Different Structural Elements in Systems

Different sections of the structure of a system in a general concept are as infrastructures or means for behaving or the discipline governing the system. With this view, some parts of structure have basic role in the system and without it, the desired system can not be existed. Also, some other parts of the system are for behavior and discipline of the system.

Body System

Organs related with discipline of the body
(brain and senses like eye, ear and other)

Structural organs   Behavioral organs
(skeleton, digestion and other)   (muscles, hands, foots and other)

Hierarchy of body organs

Government System

Organizations related with the discipline in the country
(president\prime minister, monitoring and judicial organizations, police, army and other)

Structural organizations   Behavioral organizations
(state\cabinet, ministries, parliament and other)   (executive organizations and other)

Hierarchy of governmental organizations

System in Transcendental Interpretation

When there is talking about the behavior and discipline, inevitably there is talking about the goal.

Discipline in the structure and behavior usually is meaningful when there is a goal as a destination to achieve by behaving. With this interpretation, survival of systems in order to achieve the goals needs the survival of structure with its order in space and continuing of behavior with its rule in time.

In systems, protecting the structural order and controlling the behavioral rule becomes possible by internal (due to the system itself) or external (due to the system environment)
"structural order protection and behavioral rule control mechanism".

As a more complete concept respect to the two previous concepts of the system, "system is a certain ordered structure with a ruled behavior that in the form of governing discipline in the whole of it, seeks to achieve a certain goal".

This system, besides the static and dynamic systems, can be called as "organic system".

Any system, in order to remain as it is, in its least condition has a "structure protection mechanism".

Example

a piece of stone if wants to remain as a piece of stone as it is, naturally resists against the breaking and collapsing.

Also, any system in order to continue its behavior, in addition to "structure protection mechanism" also has its own "behavior control mechanism".

Example

an ordinary door if wants to remain a door as it is, while naturally resists against breaking and collapsing, its open and close as its behavior is also controlled by the passer and the hinges.

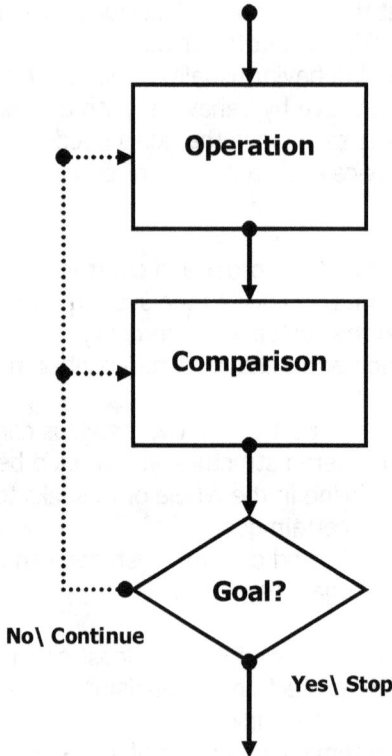

```
                    ●
                    │
                    ▼
          ┌─────────────────┐
    ●┄┄┄┄>│    Operation    │
    ┊     └─────────────────┘
    ┊              │
    ┊              ▼
    ┊     ┌─────────────────┐
    ●┄┄┄┄>│   Comparison    │
    ┊     └─────────────────┘
    ┊              │
    ┊              ▼
    ┊           ◇ Goal? ◇
    ●┄┄┄┄┄┄┄┄┄┄┄
   No\ Continue    │
                   ▼
              Yes\ Stop
                   │
                   ▼
```

In a transcendental system, the purpose of most of the behaviors is to achieve a desired goal, state or condition. For this purpose usually there is a structure and behaves in a way to achieve the goal. With any operation in the form of the behavior, current state always compared with the desired goal, state or condition in order to determine if it is achieved or not. If achieved, the work is completed. If not achieved, operations and comparisons are continued. For example, filling a cup of coffee, so that not the coffee to overflow from the cup.

Basic actions in daily activities

Analysis of the daily activities reveals that most of them are one or a combination of the three basic actions of "sequence", "selection" and "repetition". Examples of these basic actions in daily behavior are as follows:

Sequence
Doing the actions one after the another with certainty
like
cooking food, eating food and washing dishes

Selection
Evaluating the conditions and choosing one of the two or more possible ways to continue the actions
like
evaluating some thing for buying and buying or not buying it

Repetition
Doing again and again stepwise actions in order to achieve to a certain goal
like
participating in class and passing stepwise exams in order to complete the course

General types of behavior in systems

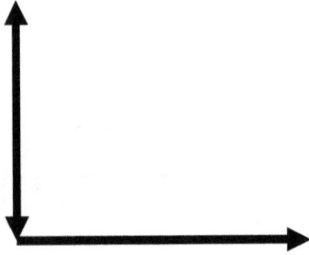

silence and no behavior
0

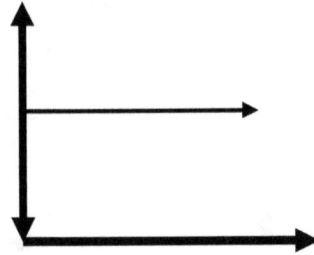

uniform behavior
1

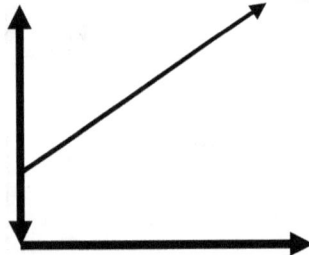

linear growth
2

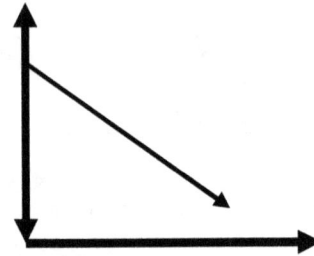

linear destruction
3

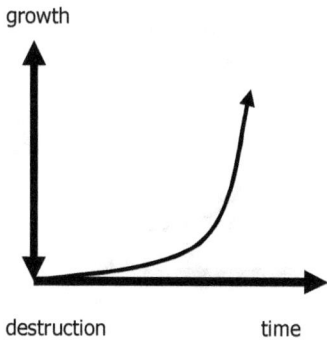

growth

destruction   time

exponential growth
4

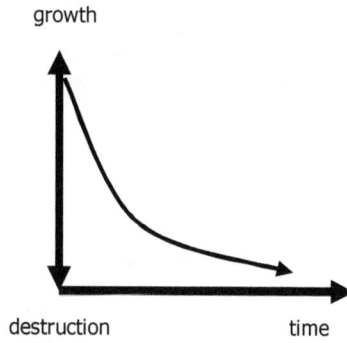

growth

destruction   time

exponential destruction
5

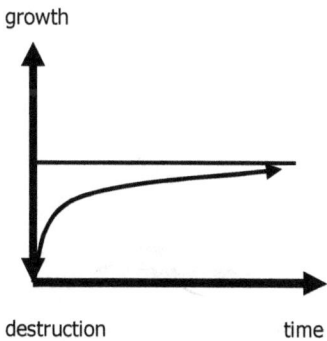

growth

destruction   time

goal seeking with increase in
growth
6

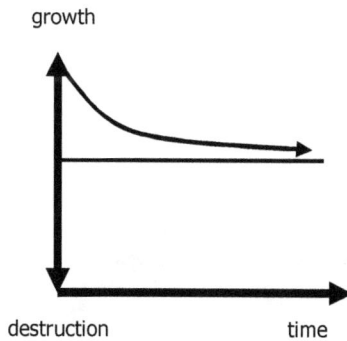

growth

destruction   time

goal seeking with decrease in
growth
7

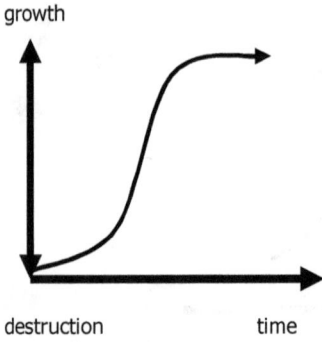

growth

destruction   time

**growth then delay and change
direction
(S-shaped)
8**

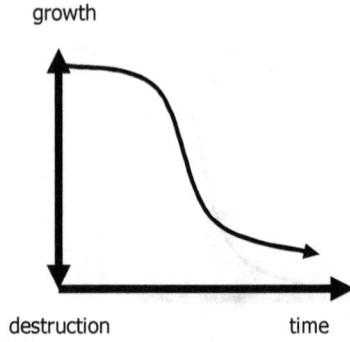

growth

destruction   time

**destruction then delay and
change direction
(inverted S-shaped)
9**

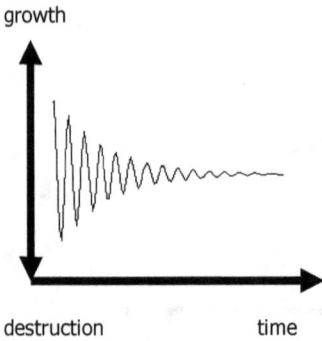

growth

destruction   time

**regular oscillating behavior
with decrease in oscillation
amplitude
10**

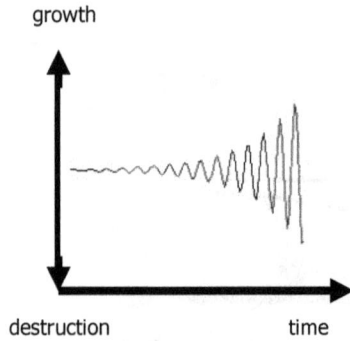

growth

destruction   time

**regular oscillating behavior
with increase in oscillation
amplitude
11**

growth

destruction   time

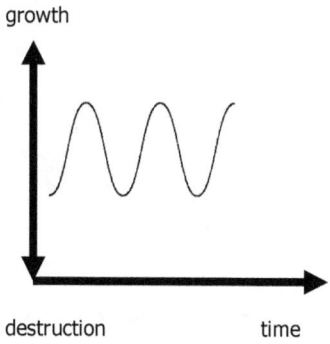

**regular oscillating behavior
with specified oscillation
amplitude
12**

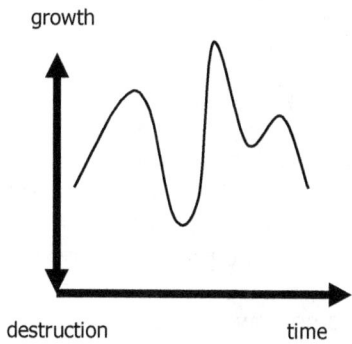

growth

destruction   time

**irregular oscillating behavior
with unspecified oscillation
amplitude
13**

Life Cycle in Systems

Any system from structural, behavioral and disciplinal views naturally has a
life cycle in the form of coming into existence (birth), applications or
functioning due to the goals (living) and going into inexistence (death).
This is done in the form of "growth" or "destruction" and finally "evolution"
or "distinction".

Birth ⟶ Living ⟶ Death

Coming into
Existence ➡ Growth ➡ Evolution ➡ Destruction ➡ Distinction ➡ Going into
Inexistence

Stability in Systems

Discipline in systems, causes while the system is faced with some aspects of constancy or silence and variation or motion, naturally and in the form of some kind of internal, external or both the internal and external "structure protection and behavior control mechanism" in space and time always some how protects its structural order and controls its behavioral rule.

The outcome of this is "stability" in system.

While constancy and variation both form the face of the world, nevertheless, systems naturally tend to be in a desirable and persistent situation that is called "stable state".

Total stability in systems is possible in the following conditions:
- "Structural balance": Minimum disorder or maximum order in the matter based structure that is possible by protecting the order necessary for the structure to exist.
- "Behavioral equilibrium": Minimum misrule or maximum rule in the energy based behavior that is possible by controlling the rule necessary for the behavior to continue.
- "Disciplinal certainty": Minimum uncertainty or maximum certainty in the information based discipline that is possible by getting, processing and using information for certainty of protecting the order necessary for the structure to exist and controlling the rule necessary for the behavior to continue.

"Order and disorder" in the structure, "rule and misrule" in the behavior and "certainty and uncertainty" in the discipline of systems is interpreted by the "entropy" scale as a reverse scale (between 0 and 1). In other words, less entropy (near to zero) is good and more entropy (near to one) is bad.

"Non-freedom" (determinism)
and
"freedom" (non-determinism)
in systems

Order and rule due to the discipline in systems, in various cases have "non-freedom" and "freedom" aspects as follows:

- Non-freedom aspect: With dominance of external protecting aspect and usually due to the governing of matter or mass and energy or force in the form of some kind of field or environment that deterministically encompasses the system.
- Freedom aspect: With dominance of internal controlling aspect and usually due to the information interchange with the environment that the system non-deterministically exists in it.
- Both non-freedom and freedom aspects: With equivalence between some degrees of deterministic external protection and non-deterministic internal control aspects.

In the above mentioned spectrum of the objects in nature, from the beginning to the end of spectrum, determinism or non-freedom aspects in the discipline decreases and non-determinism or freedom aspects increases and information plays an important role in it.
In systems that besides matter and energy also can use information, there are more aspects of freedom.

System and Systems Thinking

System view on objects and their related events requires that in dealing with any objective or subjective thing, while paying attention on its parts and relations among as the whole including the appearance or structure of the thing, its function including the interior or behavior and discipline including the structural order and behavioral rule governing the thing are also considered.

This is done under the concept of system and systems thinking and by the use of a set of concepts and methods in the case of objects and related events and governing orders and rules and structure, behavior and discipline in the whole of it.

System and systems thinking are the general title of the concepts and tools that are used to explain the world.
This type of approach to the world is so natural that as we saw, the trace and effect of it can be seen in the natural languages as one the most common and available tools to explain the world.

System and systems thinking are as the combination of the two old trends in dealing with the world.
These two old trends are
"reductionism" (emphasizing on the parts forming an object)
and
"holism" (emphasizing on the whole due to the relations among the parts forming an object).

Reductionism by emphasizing on structural elements led to some kind of "structuralism".
Holism by emphasizing on the behavioral wholeness, led to some kind of "behaviorism".

Reductionism can be called as
"systemic approach" or "order based approach".
Holism can be called as "systematic approach" or "rule based approach".

The combination of these two approaches including the combination of structure and behavior also may be introduced in the form of "organization" (from structural order and behavioral rule view) and "management" (from structural order protection and behavioral rule control view).

This kind of approach that implies the organizational and managerial aspects can be called as "systems approach" or "systems thinking".

Today, in many cases, "lack of system" means the "lack of order in the structure, rule in the behavior and mechanism for protecting the structure and structural order and controlling the behavior and behavioral rule".

For this reason, one of important aspects of the era we live in, is organization and management of the objects and related events
and
the two subjects of organization and management are among the most important subjects in the contemporary world and especially in the field of systems.

Structure and Static Aspects Presentation
Picture Diagram

In "picture" diagram, each element of the system is displayed by an appropriate "icon" or "symbol" in the form of a small picture that has similarity or relation with the element and the name of the element is written besides the related icon or symbol. Also for displaying the relations among the elements, other appropriate icons and symbols may be used between two elements. In this case, relation between two elements that are besides each other in the diagram can be shown in the space between the elements.

In the case of hard physical systems, it is ideal that the diagram representing the system, to be similar to the physical structure of the system as more as possible. For example the line drawings of different devices or home appliances (like home televisions) that come in the installation manuals of the devices, are good examples of these diagrams. In these diagrams, the structural elements and relations or links among them are presented in order to present the whole device and in general display the physical structures of the related systems. In the case of logical understandable systems, pre-agreed icons or symbols may be used.

In picture diagrams, various signs may be used for modeling the system of interest. Some of them are as below:
- "Icon": "Direct" or "indirect" displaying of the objects in the form of a picture of them with some kind of "relation" with the objects of interest. Like using the shape of trash can for displaying the deleted files or the shape of a magnifier or binoculars for displaying the search of information in the computer or other similar shapes in traffic signs.
- "Symbol": "Arbitrary" or "optional" displaying of the objects in the form of a picture with some kind of "pre-agreed concept" in relation with the objects of interest. Like the use of red triangle for displaying warning in the computer or other similar shapes in traffic signs.

To draw picture diagrams for displaying the structure and structural order of the systems has no certain rule. It must be said that picture diagrams have more artistic aspects than technical aspects. Hence, for drawing picture diagrams, in addition to technical information, some kind of artistry is needed. In practice, for drawing pictorial diagrams, some sets of symbolic pictures may be gathered and classified for different uses. In this case, each set of symbols can be used for a specific kind of system and each symbol in a set for a specific element in the system as a standard. Different sets of these symbols may be gathered and kept for using in different systems. For drawing picture diagrams also there are many software tools.

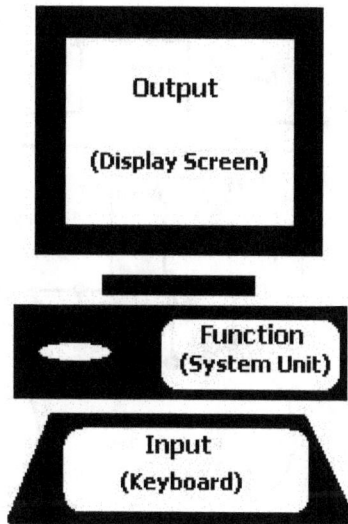

An ordinary computer is a good example of a system in its complete concept. In computer, the three steps of "Input – Function – Output" are done by specific structural elements. Getting in the primary data (as input) is done by "keyboard", processing data as the "function" of system by "system unit" and putting out the results by "display screen". Input unit is like eye for seeing or ear for hearing information, system unit is as brain and paper notes for processing and keeping information and output unit is like language for speaking or hand for writing information in humans. Previous man made machines only used or processed matter and energy, but computer can use or process information as the third basic resource.

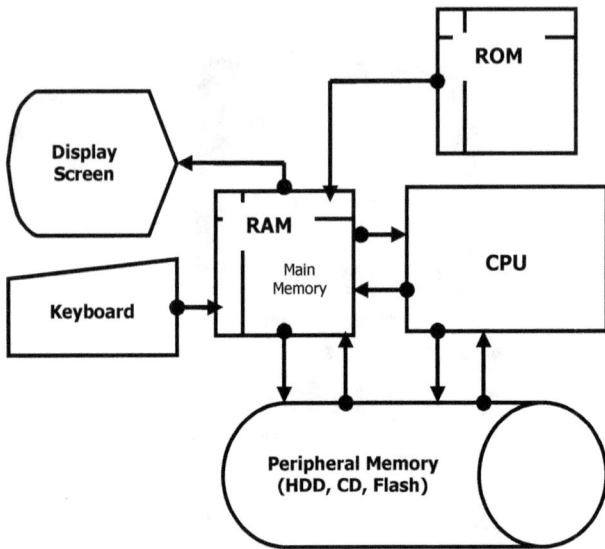

RAM: Random Access Memory
ROM: Read Only Memory
CPU: Central Processing Unit
HDD: Hard Disk Drive
CD: Compact Disk

In this "picture diagram" the whole structure of a computer in general is shown as a system. As it is seen, each structural element of computer is displayed by a specific pictorial symbol that reminds its shape or has a relation with it. In this diagram, the "system unit" in the previous diagram, is consisted of ROM, main memory or RAM, processor or CPU and different types of peripheral memory\ storage. ROM is like the battery in a car and is used for starting the computer. Main memory or RAM is like the blackboard in a class that each empty corner of it is as a working or temporary memory for processing inputs by CPU. Peripheral memory is like the students' paper note books in a class and unlike the blackboard as temporary memory, is permanent memory. Programs and data are entered to the computer by the input unit and stored in the peripheral memory. Programs by coming into the main memory as working memory are executed by processor and by the use of input data, produce the output results. Results sent to the computer users by the output unit or stored on peripheral memory for further uses.

Block Diagram

In block diagram, each element of the system is displayed by a "block" or "box" in the form of a rectangle and the name of the element is written inside it. Like picture diagrams, also in block diagrams relations between the elements may be displayed in the space between the blocks of related elements. Of course, in practice this type of displaying the relations between the elements may be restricted to the physical connections between the elements. Usually, interchange of the main and basic resources including matter, energy and information between the elements, may be displayed in the behavior of system and flows from one to another.

Because the block diagrams include some kind of abstraction, to draw it, more or less has a rule. The first abstraction is that, in block diagrams structural elements apart from their specific properties, are displayed in the form of a rectangle. This abstraction and uses of it, leads to some kind of rule in drawing the block diagrams in the case of different systems.

To draw picture and block diagrams for displaying the structure and structural order of systems, following steps may be taken:

- Determining structural elements of the system
- Determining the exact name of each structural element
- Determining the structural order of elements from the view of being besides each other
- Determining the relation between an element with other elements according to the structural order of the whole system
- Determining the exact title of each relation between two elements
- Determining the core, central or main elements
- Drawing the total scheme of the diagram
- Determining the place of core, central or main elements in the total scheme of the diagram
- Determining the arrangement of other elements around the core, central or main elements
- Selecting the appropriate "pictorial sign" (icon or symbol) or "rectangle" (block or box), good for each type of element
- Placement of the selected pictorial signs or rectangles in the total scheme of the diagram
- Writing the name of the elements besides the pictorial signs or inside the rectangles representing the elements
- Writing the titles of the relations between the elements near the pictorial signs or rectangles representing the elements
- Reviewing the whole diagram and removing or modifying the cases that may lead to misunderstandings

In structural diagrams, for displaying the belonging of the elements to the system, it is possible to bind the whole elements inside an area with a closed dotted line as its boundary. Like this, also the elements of sub-systems under a main system can be bound by their own closed dotted lines. In this case, the bigger closed dotted line as the boundary of the main system includes all of them. Indeed, each closed dotted line is the boundary of the related system. Also, inside of the bigger closed dotted line represents the environment of the sub-systems under the main system and outside of it represents the environment of the main system. For distinguishing the boundaries from each other, different forms of dotted lines may be used.

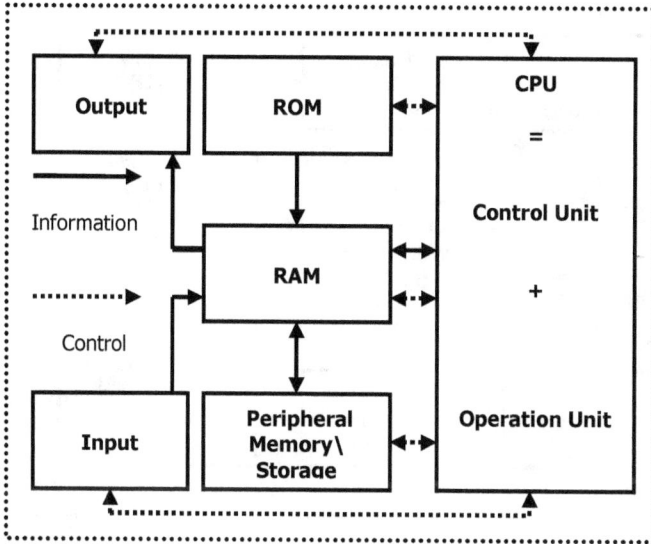

In this "block diagram", structure of a computer is displayed. CPU is consisted of two basic units of "control unit" and "operation unit". Full line arrows display the flow of data or information and dotted line arrows display the flow of control signals. In computer terminology, "data" and "information" have similar meanings, but "data is raw information" that is put in to the computer and "information is processed data" that is put out from the computer.

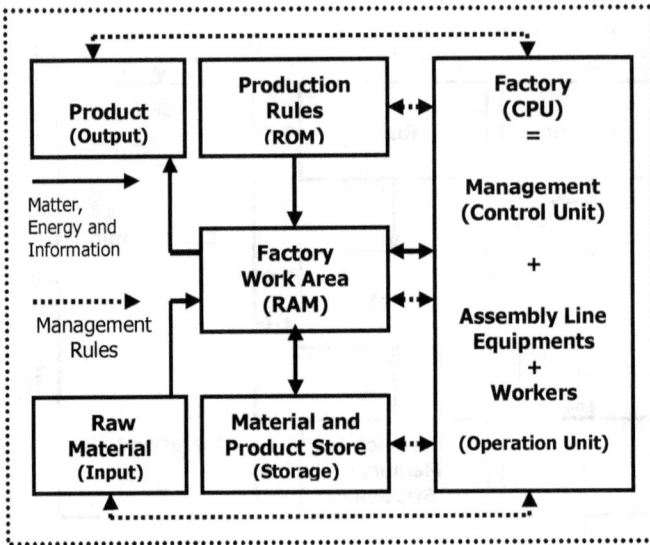

In this "block diagram", units forming a factory are displayed. Attention on the functions of computer and factory and comparing this block diagram with the block diagram of a computer presented earlier, shows the similarities between them. It must be said that according to the systems view, both are as "system". In factory, raw materials come into the working area of the factory and according to the production rules, by the equipments and workers in the assembly line converted to products. Also in computer, programs and data come into the main memory and executed or processed by processing unit or CPU. Production rules are like the contents of ROM, working area of factory is like RAM and inventory or store of raw materials and products in factory is like peripheral memory or storage in computer. The role of management and assembly line in factory is like the role of CPU (including control and operation units) in computer. Full line arrows display the flow of matter, energy and information and dotted line arrows display the flow of management rules.

Behavior and Dynamic Aspects Presentation

Behavior of a system in general, includes sets of events, flows of states or steps with constancy or silence and variation or motion, actions and reactions, requests and responses, flows of causes and effects, stocks and flows, functions, processes including input - processing\ functioning - output, life and its cycles. Behavior also includes following up the goals, purposes and ideals, struggles to create and maintain balance in structure, equilibrium in behavior and certainty in discipline, adaptation with the environment, learning from the environment and evolving more from the structural, behavioral and disciplinal views for better functioning in order to achieve more goals, purposes and ideals in future.

Behavior and behavioral dynamic aspects in systems also may be displayed by the use of what we said above about displaying the structure. The difference is that, because of the behavior as the main subject in the diagram, the purpose is to display the flows of variation and motion or behavior of the system. In this case, the structure may be summarized to its minimum so that only to display or support the desired behavior.

In addition to this, for displaying behavioral aspects in systems, also there are many kinds of specific diagrams. Since all of these diagrams display the "flow of operations" in systems, as we mentioned before, all have the general title of "flow diagrams". In the following sections, some of the general and common flow diagrams for displaying the behavior and behavioral rule in systems are explained.

Flow Chart

"Flow chart" is one of the oldest, most known and common among the flow diagrams. In flow chart, boxes and arrows are used to display the behavior in systems. In flow chart actions in the form of behavior are displayed by specific boxes or blocks for specific actions and linked to each other by arrows. Also the title of each is written inside it. In flow chart, operations are sequential in time and in some steps may be simultaneous.

A general version of flow chart that can be used to display the behavior in most systems includes the six symbols for "start\ stop", "operation", "decision making\ control", "input\ output", "sub-system under main system" and "connector". The symbol for "start\ stop" is an ellipse and the word "start" or "stop" is written inside it. For displaying "operation" a rectangle, "decision making\ control" a diamond and "input\ output" a parallelogram is used.

A sub-system under a main system is displayed by a rectangle with double right and left sides. All the symbols are linked to each other by arrows to show the flow of behavior from start to stop. To continue the diagram in the next page, first in the current page it is ended to a circle with a letter in it and then this circle is repeated in the next page for continuing the diagram.

"Start" has out-going and not any in-coming arrow. Unlike start, "stop" has in-coming and not any out-going arrow. Connector in the ending point has in-coming and in the start to continue point has out-going arrow. Other symbols that are used in the body of diagram can have in-coming and out-going arrows. Regularly, diamond representing the decision is in the form of a question or condition and has one or more in-coming arrows from one corner and three other out-going arrows from three other corners as "positive" (for "yes" case), "zero" (for neuter or otherwise case) and "negative" (for "no" case).

The question or condition also can have multiple responses or out-goings in the form of "multiple cases and otherwise" including "case-1" up to "case-n" and "otherwise". So, decision diamond can have one in-coming and n+1 out-goings for n "case" and 1 for "otherwise". For displaying the out-going cases, a horizontal line is used. For doing this, first form the decision diamond one arrow is drawn to this horizontal line. Then, from this line n+1 arrows are drown for n cases and 1 for otherwise case.

Operations or decisions can be classified for example according to the location and date (space and time) of performing them. Classification helps the readers better understand the operations and decisions and the space and time of them. For classification, a big table may be used with the titles of its columns as the titles of classes, for example the locations or spaces of operations and decisions. Symbols according to their classes are placed in the columns of the table and linked to each other by arrows. In this case, the vertical lines of table will act as "separator" of different classes.

Like what told about the columns of the table in the above, the rows of the table may be used for classification of the date or time of operations and decisions. Now each cell of the table will act as a place for a symbol of operation or decision in a specified space and time. If the description of operations and decisions can not be written inside the symbols, then they can be coded and by adding a new column to the table, can be written in this column by the aid of code of each symbol.

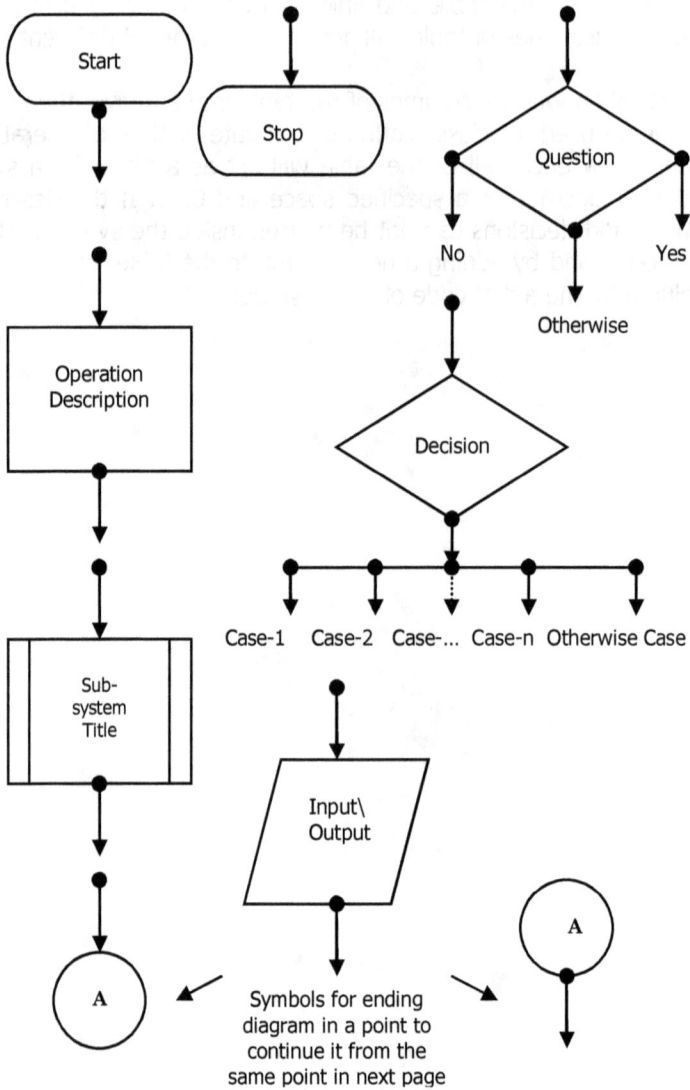

Start

Stop

Question

No

Yes

Otherwise

Operation
Description

Decision

Sub-
system
Title

Case-1 Case-2 Case-... Case-n Otherwise Case

Input\
Output

A

A

Symbols for ending
diagram in a point to
continue it from the
same point in next page

Main symbols used in flow chart to display behavior in systems

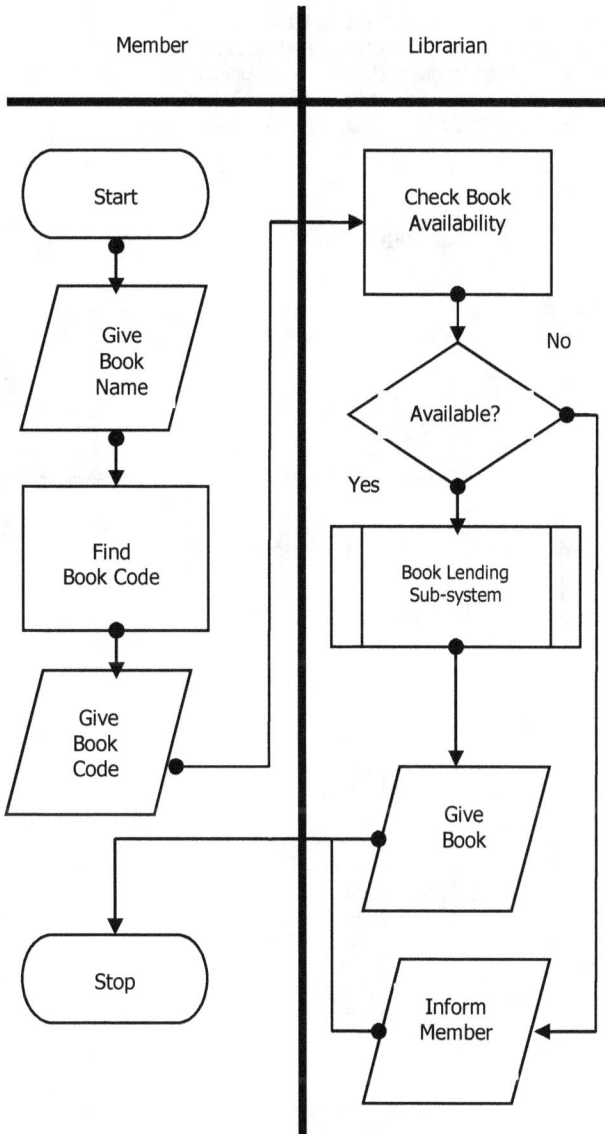

Member	Librarian

Start

Give Book Name

Find Book Code

Give Book Code

Stop

Check Book Availability

Available?

No

Yes

Book Lending Sub-system

Give Book

Inform Member

Flow chart for borrowing a book from library in traditional method

Displaying Basic Actions in Daily Activities

Following is the display of "sequence", "selection" and "repetition" basic actions in daily activities, by the example of "going to buying". Actions are displayed first by the use of flow chart and then by its equivalent in the form of "box diagram" that is known by the name of its inventors as "NSD" or "Nassi- Shneiderman" diagram.

Sequence

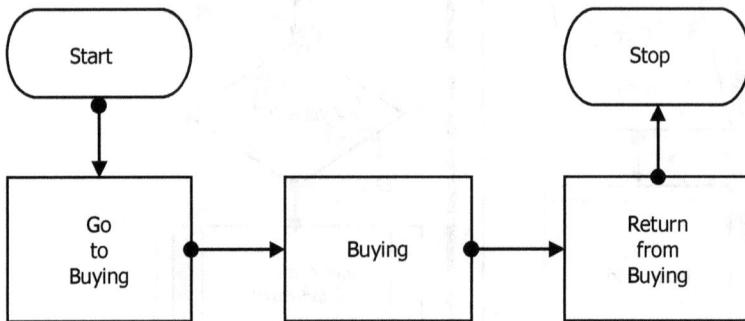

Selection

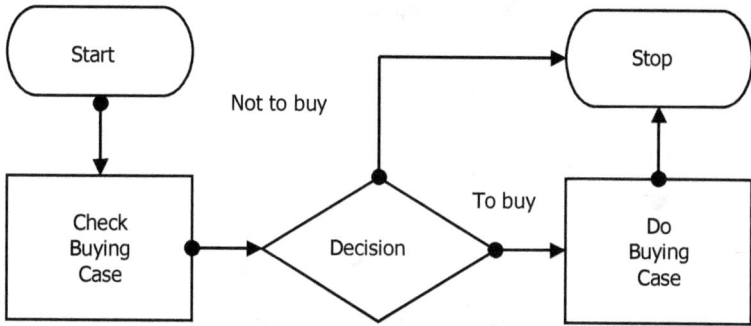

Repetition

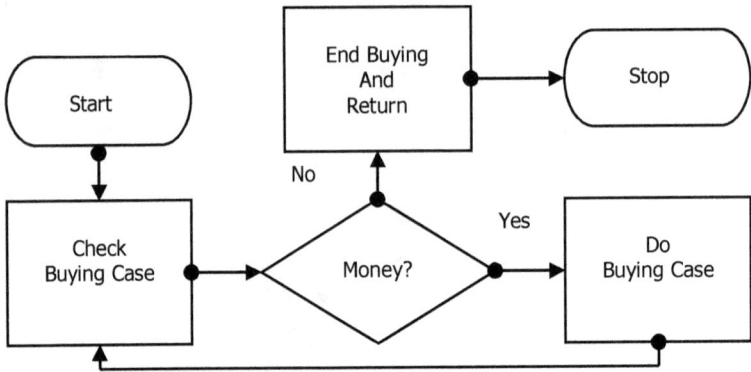

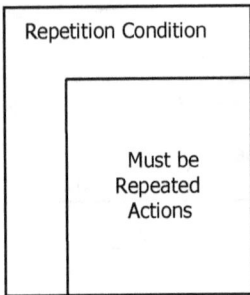

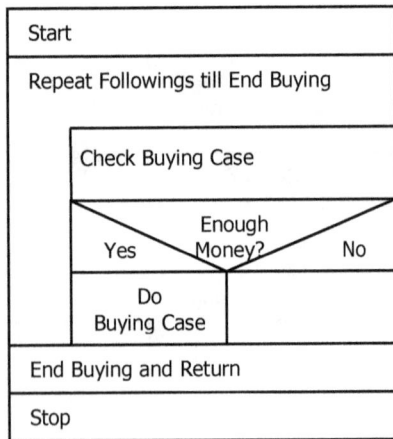

Activity Diagram

"Activity diagram" as a new form of flow chart, is a new tool for modeling the behavior of systems. Activity diagram has a certain "start" and "stop" point and includes "activity" and "decision". In activity diagram, "start" is displayed by a filled small circle and "stop" by a bigger circle with a filled small circle (like start) inside it. Like flow chart, also in activity diagram, activities displayed by rectangle and decisions by diamond. Activity diagram is one the UML (Unified Modeling Language) diagrams that is mainly used in computer software design. Activity diagram respect to flow chart has fewer symbols and includes some modifications.

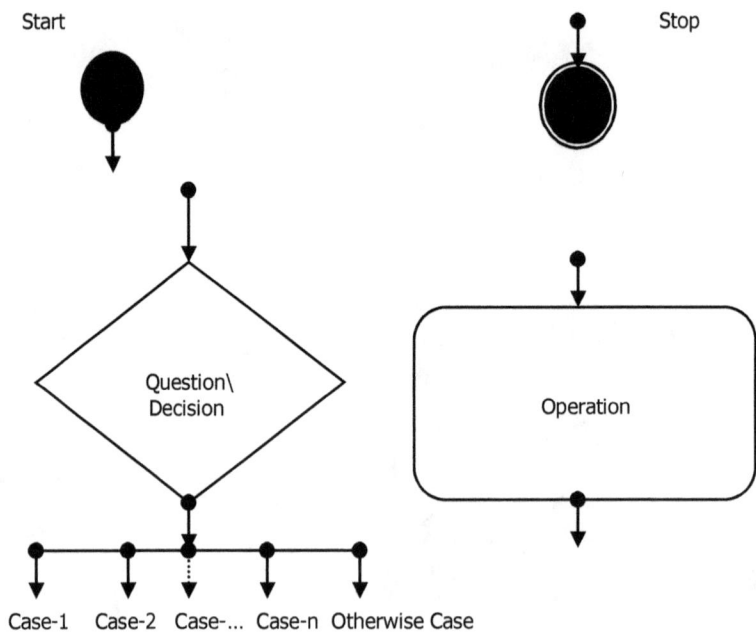

Start

Stop

Question\
Decision

Operation

Case-1   Case-2   Case-...   Case-n   Otherwise Case

Main symbols used in activity diagram to display behavior in systems

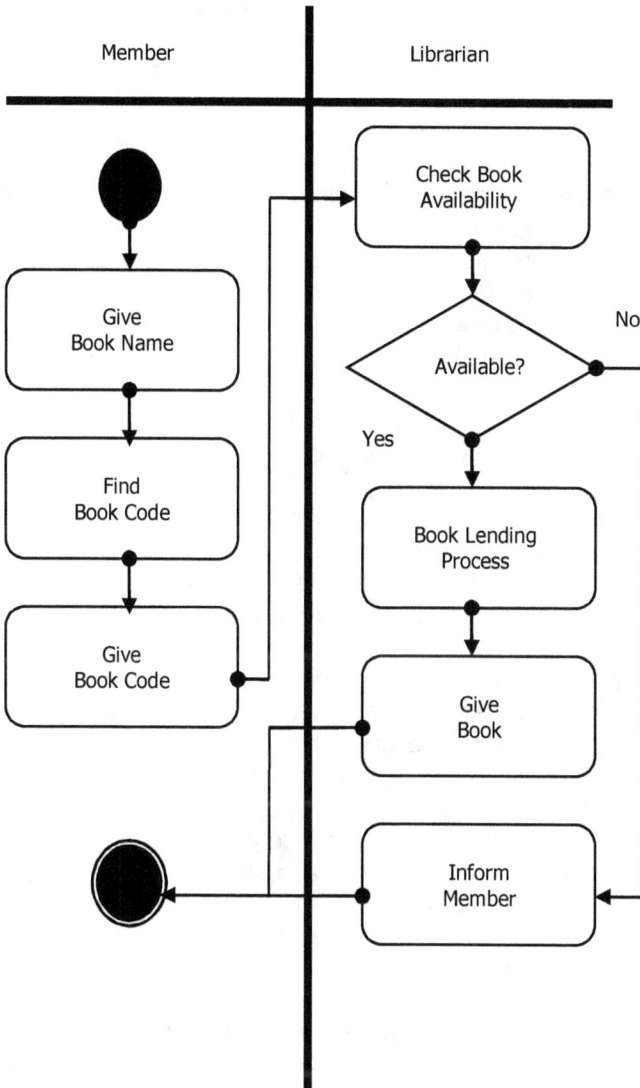

Activity diagram for borrowing a book from library in traditional method

State Chart

"State" is a certain situation in a certain time in a system behavior in that there is a certain structural order and a certain behavioral rule and the situation is important from the system researcher's view. State is also a certain and meaningful combination of system properties in a certain time. In this case, the state is determined with the values of the properties in that specific time.

Behavior of some systems is displayable in finite and limited steps or states by a "state chart". "State chart" also is named as "State Transition Diagram = STD". It can be said that system "transits" from one state to the other state during its behavior. State chart as the "system states model" is a display of all possible states in the behavior of system. For drawing the state chart of a system, first all the possible and desired states in the behavior of system and the order of states must be determined.

For example, the home audio or video devices or washing machines are good examples of this kind of systems. Indeed, states of these devices from their users' view can be supposed by the buttons that are used for different actions. From state view, static systems are "single state" and dynamic systems are at least "two state" like "on" and "off" or "multi state" like other complex dynamic systems. The situation of each state of system can be seen in some way in the structure of the system.

States usually are transit able form one to other by specific actions. The action that transits one state to another state is displayed by an arrow from the previous state to the next state. The state chart is formed by connecting the various states to each other in this way. In a system with a state based behavioral pattern, "state transition" means going from one state to the other state in order to continue the behavior in the new state. In this case, system is transited from a state as "origination" to another state as "destination". In any state transition, some "pre-condition" (the condition that must be true before transition) and some "post-condition" (the condition that must be true after transition) may be necessary.

Transition from an origination to a destination state needs the establishment of pre-condition in the origination and post-condition in destination state. Any state transition usually has its rule that is called "transition rule". The rule may contain the ending of an action, occurrence of an event or attainment of certain conditions in the state. Transition from an origination to a destination state occurs according to this rule.

In state chart, each state is presented by a rectangle. Rectangles representing the states in the behavior of system, connected to each other by arrows representing the actions must be done in order to transit the system from one state to the other. For displaying start and stop, the same symbols of the activity diagram are used. Start symbol is connected to the first starting rectangle and the last ending rectangle is connected to the stop symbol by an arrow.

In each certain state of the system behavior, different actions may be done by the system. The most important actions are as follows:
- "State Entry Action": The action that the system must do when enters the state.
- "State Stop Action": The action that the system must do when stops in the state. This kind of actions can be classified in two major classes as "foreground action" (direct) and "background action" (indirect).
- "State Exit Action": The action that the system must do when exits the state.

Also according to the situation in the flow of behavior, states can be classified in the most general classes as bellow:
- "Initial\ Start State": State with no previous state before it, like the "on" state in an "on\ off" able device.
- "Mid State": State like each of the states in the behavior of a system with previous state before and next state after it.
- "Current State": A state that the system when behaving, "now" is in it and it may be important from the researcher's view in a specific moment of time.
- "Stand-by State": State with stop while being ready to behave.
- "Pause State": State like the "Stand-by State" in that the behavior of system temporarily suspended for ensuring some conditions and after that system can continue to behave. Like suspending the speaking in order to drink water by the speaker.
- "Terminal\ Stop State": State with no next state after it, like the "off" state in an "on\ off" able device.

Finite State Machine\ Model

Like state chart, "Finite State Machine\ Model = FSM", is another tool or method for displaying the behavioral patterns in systems. Finite state machine is a virtual machine with a behavioral pattern consisted of limited and specified states for doing limited and specified actions that is used for displaying the behavior of complex systems. Finite state machine has inputs, memory and outputs. The behavior of a finite state machine starts from a primary state called "initial\ start state". Then, by passing from "mid states" each as "current state" in a specific moment of time, reaches to a final state as "terminal\ stop state". Also here, entering to states is due the actions or events that occur.

During the behavior, finite state machine may move to inactive states as "stand-by" or "pause" states. Like state chart, finite state machine is also an abstraction of possible states that is used for modeling the behavior of systems. Also here, there are limited and specified states with actions that may be done in each state and the conditions and rules for transition from one state to another. Special finite state machines that usually are called with the name of their inventors have special rules for displaying and transition of states.

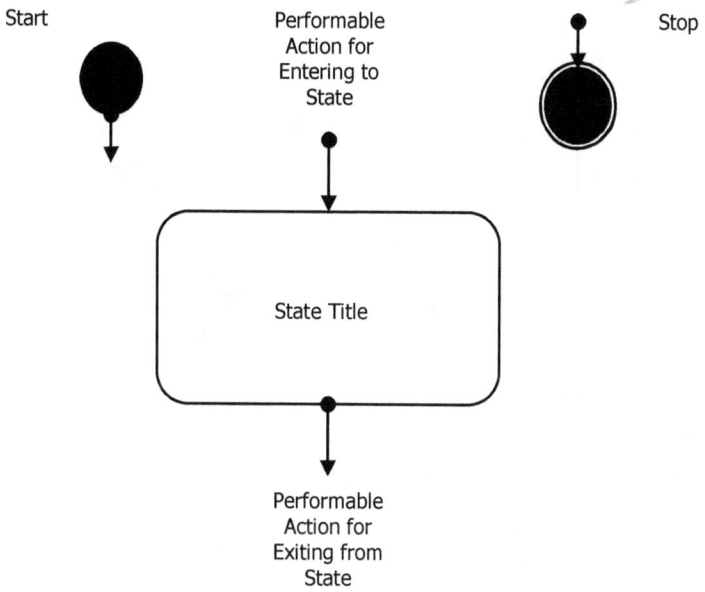

Start

Performable Action for Entering to State

Stop

State Title

Performable Action for Exiting from State

Symbols used in state chart and finite state machine for displaying behavior in systems

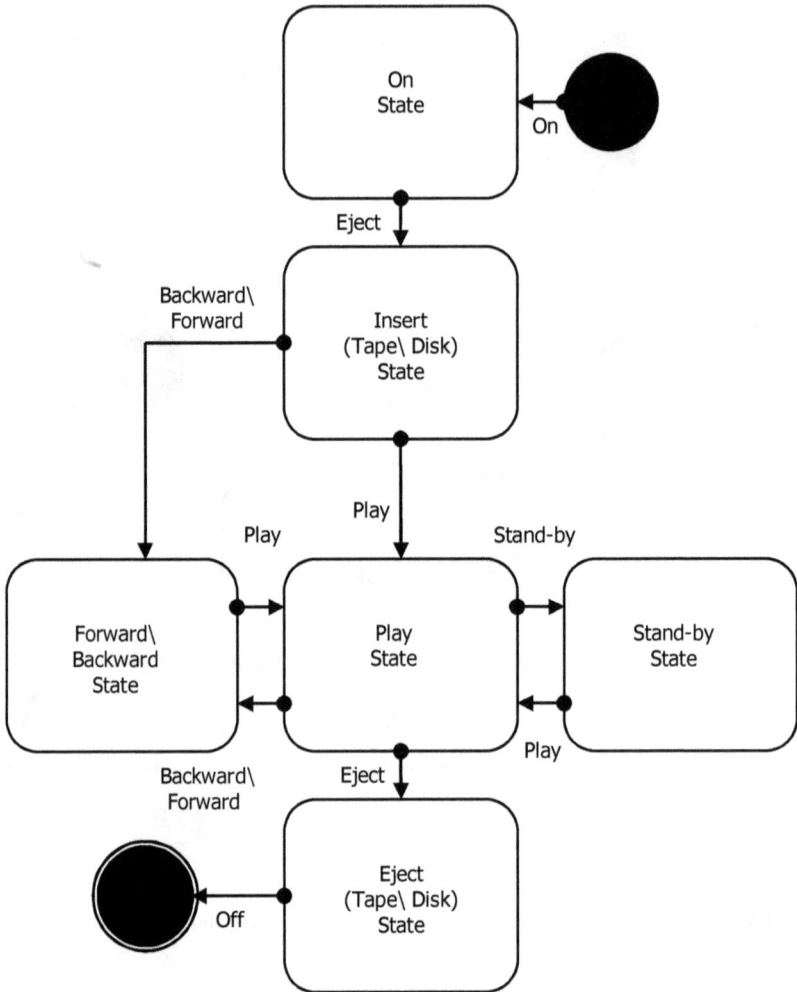

Different states and state transitions in a home audio\ video device
(Arrows as pressing the related buttons)

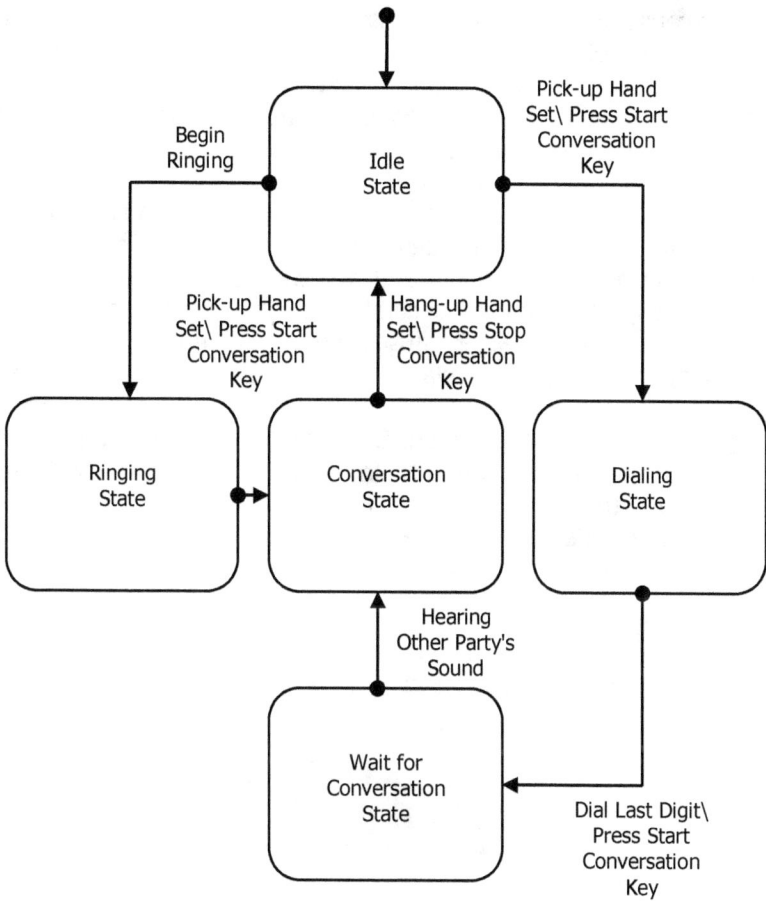

Telephone is an example of a finite state machine. In the above, a diagram of different possible states of a telephone when using it is displayed.

Cause and Effect Diagram

In "cause and effect diagram", relation between the cause and effect is displayed. As we saw, there are two types of cause and effect relations as "linear" and "circular". "Linear cause and effect diagram" is used in limited and simpler and the "circular cause and effect diagram" is used in complex cases. Linear cause and effect diagram is a simple and limited diagram in that each cause is related with its effect by an arrow in one direction. Circular cause and effect diagram is a complex diagram in that while like before the cause is related with its effect in one way, also the effect is related with its cause in another way.

Cause and effect diagrams show the flow of change from cause to effect and not show the flow of time between them. In other words, a cause and effect relation between a first thing and a second thing, does not mean "first the first thing and then the second thing" in time. But means that change in the first thing as cause, causes the change in the second thing as effect. In the cases that the relation between the two things are not clear, for displaying relation between them, it may be necessary to add a third thing as interface between them. Cause and effect diagram must be far from complexity and summarized to show the desired cause and effect relation and finally the behavioral pattern. In any case, cause and effect diagrams must be simple enough to be understood by others.

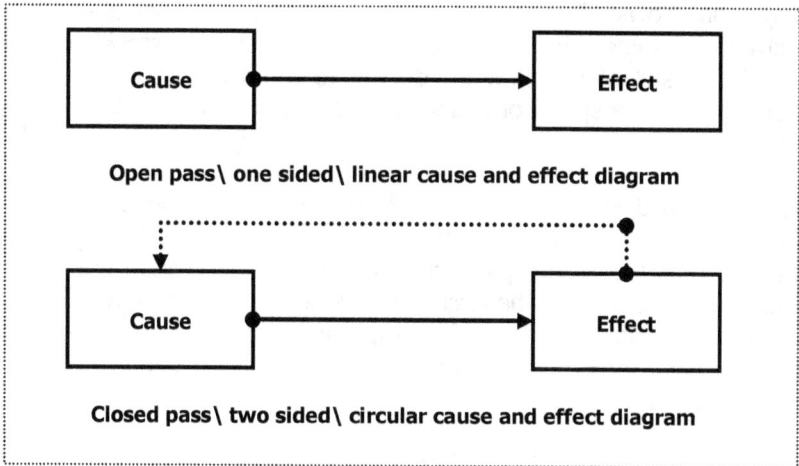

Open pass\ one sided\ linear cause and effect diagram

Closed pass\ two sided\ circular cause and effect diagram

"Cause and effect" process is one of the most ancient processes man ever has face with. According to this process, each "effect" is due to an affecter as "cause". In the past, the process was thought to be open and in one direction. But man gradually found that not only the effect is affected by the cause, but also the effect affects the cause. So, the related process instead of being open and in one way is closed and in two ways.

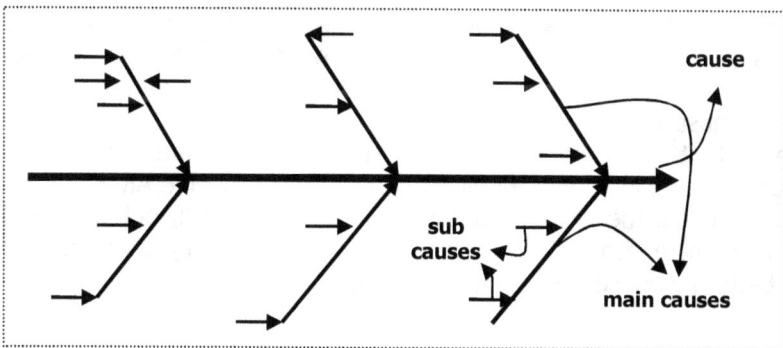

One of the cause and effect digram types is one that because of its apparent shape, is called "fish bone diagram" and is famous by the name of its inventor as "Ishikawa Diagram". In this diagram, effect is displayed by a core or central arrow and the causes leading to this effect, are displayed by other arrows around it in different classifications. One of the benefits of this diagram is the possibility of classification of causes and determining the order of causes in each class.

Stock and Flow Diagram

"Stock and flow diagram" is used to display the processes with stocking of some thing and then flowing of it to other stocks of the same kind or consuming it. In stock and flow diagram, changes on "stocks" occur by the changes on "flows". "Stock" is a source, store or container and "flow" is the stream of some thing with a certain "rate" from one stock to another stock. The result of this flow is addition to a stock by subtraction from another stock. The speed of changes in stocks is determined by the flow rates.

Stock and flow diagrams are used in forecasting the changes in processes before the real appearance of the changes. Stock and flow diagram also can be used as a base for quantitative modeling in studying the properties of the processes. Besides the circular cause and effect diagrams, stock and flow diagram may be used for displaying the relations among the quantities that vary in time. For drawing stock and flow diagrams there are special software tools.

Unlike circular cause and effect diagram, stock and flow diagram distinguishes between the different variables. Variables include "stocks" and "flows". For understanding or drawing a cause and effect diagram, the stocks and flows must be identified from each other. Stock and flow diagram respect to circular cause and effect diagram explained before, shows more information about the processes.

Of course, also the stock and flow diagram can not respond to all of the questions about the processes. For responding more questions, besides the diagrammatical presentations, also the quantitative aspects of the processes must be considered. For simplifying the problem, it may be considered that flows are continuous and constant in time. This means that the flow of quantity in the unit of time remains constant. Although this is not true for all the cases in all times in the real world, but when the cases are more and also the time is supposed to be long, then this will lead to better responses with better approximations.

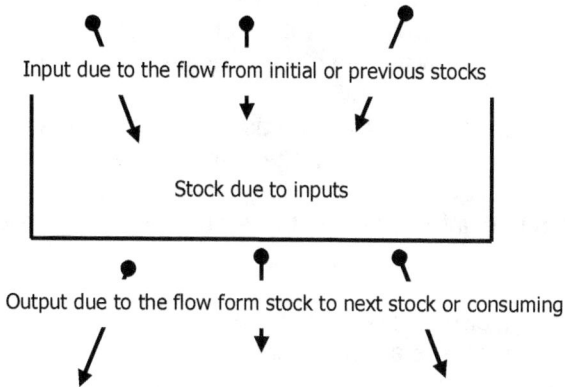

Input due to the flow from initial or previous stocks

Stock due to inputs

Output due to the flow form stock to next stock or consuming

In this diagram, the process of stock and flow that can be seen in the world around is displayed. For example, birth and death, migration in and migration out, employment and unemployment, saving and loan and in a general form, all types of inputs for storing and outputs for storing in other places or consuming, are examples of this process. Stocks display the reservoirs and flows display the amount of exchange of reservoirs form one to another in any moment.

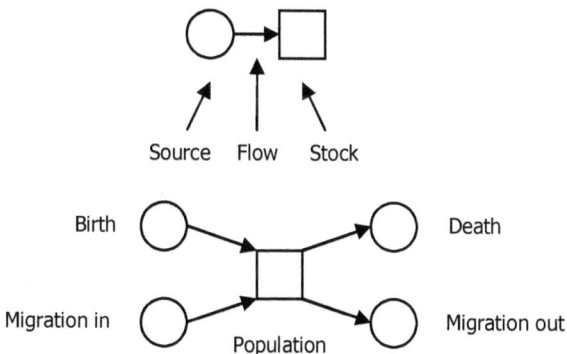

Source   Flow   Stock

Birth   Death

Migration in   Migration out
              Population

In stock and flow diagram, "flow" from "source" to "stock" is displayed by the above diagram segment. A total stock and flow diagram is displayable by connecting the above diagram segments.

Use Cases Diagram

"Use cases" diagram is a tool for modeling the user interactions of a system with system. In other words, use cases diagram is for displaying the behavior of system from the view of its users and distinguishing the interactions with system from the total behavior of system. Users of a system, interact with the system in order to achieve their own goals.

Each "use case" indeed is a certain interaction of a user or a class of users with system. During interactions with the system, users can have their own specific roles with their own specific titles. "System user\ actor" is a real or virtual character that interacts with the system in the form of different "use cases".

Use cases diagram also is used to display the sub-systems of a main system. Indeed, use case is a tool for breaking the system into sections as sub-systems. Use case shows the high frequency events in the behavior of system. Use case as a starting point in analyzing the behavior of system, indeed models the total behavior of the system. Use cases diagram is also one of the UML diagrams mainly used in software design.

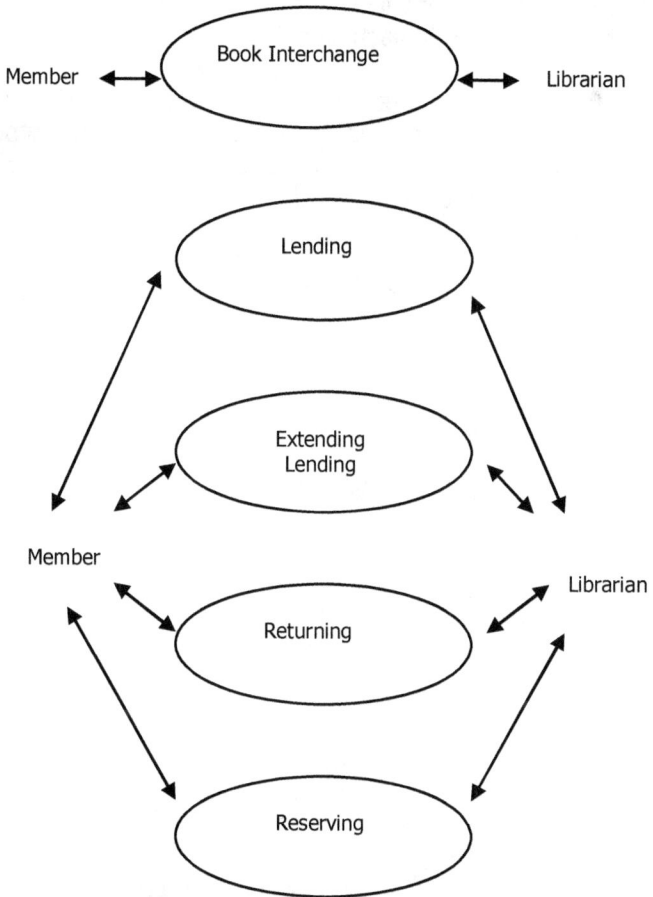

Use cases diagram displaying interchanges between member and librarian in a library
Up: Total, Down: Main Components

Process Flow Diagrams

"Process flow diagram" is used to display the flow of operations consisting of "input", "function" and "output". Processes have inputs in the form of raw and outputs in the form of processed of the three main and basic resources of "matter", "energy" and "information". A process flow diagram can be organized in three steps including "input", "function" and "output" and the main resources used in each step.

In a process, first the inputs are entered from one end. Then the pre-defined functions use inputs to produce outputs. Finally the outputs that are the results of functions on inputs are exited from the other end. In a process, some of the outputs are wanted (like products) and some others are unwanted (like wastes).

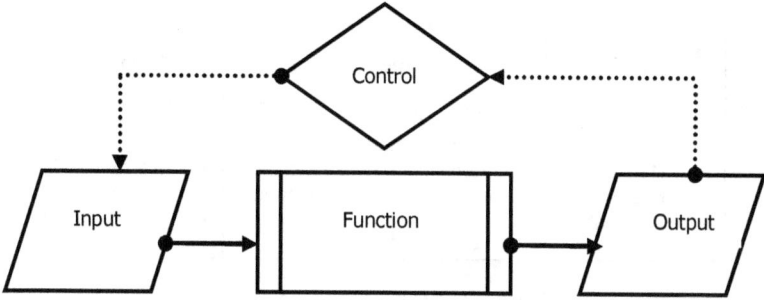

A process flow in general

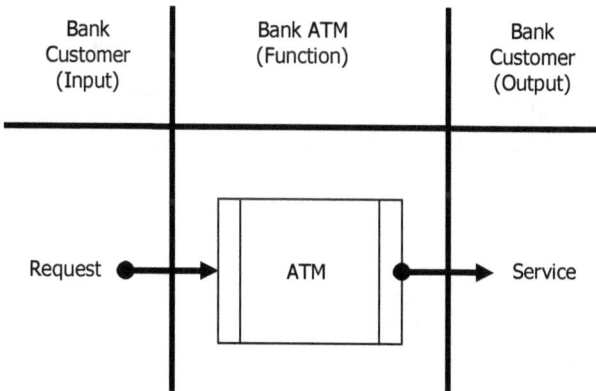

Total interaction of a bank customer with bank ATM

Input, Process and Output		
Input	**Function**	**Output**
Raw • Matter • Energy • Information	• All the actions on inputs to produce outputs • Combination of sequence, selection and repetition in behavior	Processed • Matter • Energy • Information

Input, Function and Output in a process in general

System Title		
Input	**Function**	**Output**

To describe the inputs, processes and outputs in a system, this form that contains the titles and short descriptions about each of the mentioned cases, can be used.

Discipline and Protection and Control Aspects Presentation

Disciplinal mechanism in systems has two main duties of protecting the structural order and controlling the behavioral rule. In consequence, the diagram that displays the mechanism can have two aspects of "order protection" and "rule control". Of course, structural diagrams display the order in the structure and behavioral diagrams display the rule in the behavior of the system in some way. At the first glance, a combination of structural and behavioral diagrams in the case of a system, also can display the discipline in the system in some way.

But what is important here in the case of a diagram from disciplinal view in a system, is the addition of the flow of "information" besides the flow of "matter" and "energy" and formation of "closed feedback loops of information" in diagram. In a combinational diagram of structure and behavior, when the flow of information is added, then the diagram will be useful also from disciplinal view.

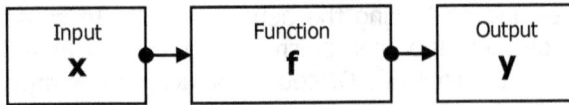

```
┌──────────┐      ┌──────────┐      ┌──────────┐
│  Input   │─────▶│ Function │─────▶│  Output  │
│    x     │      │    f     │      │    y     │
└──────────┘      └──────────┘      └──────────┘
```

$$y=f(x)$$

Systems without protection and control mechanism
In this case (like in solids and for example a piece of stone or an ordinary door), protecting structural order and controlling behavioral rule is possible by the physical (or natural) strength of the elements and relations among or an external agent (like the passer of the way or the door).

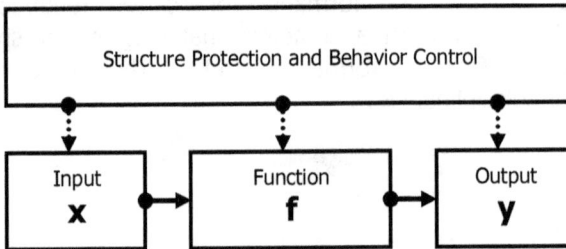

```
┌──────────────────────────────────────────────┐
│      Structure Protection and Behavior Control │
└──────────────────────────────────────────────┘
      ┆            ┆             ┆
      ▼            ▼             ▼
┌──────────┐ ┌──────────┐ ┌──────────┐
│  Input   │ │ Function │ │  Output  │
│    x     │ │    f     │ │    y     │
└──────────┘ └──────────┘ └──────────┘
```

$$y=f(x)$$

Systems with one sided\ open protection and control mechanism
In this case (like in plants and for example a vase of flower), mechanism due to an external agent, has external aspect and without internal feedback and flexibility. So, system always can not adapt to the conditions, except by physical (or natural) strength of the elements and relations among or the external agent involvement.

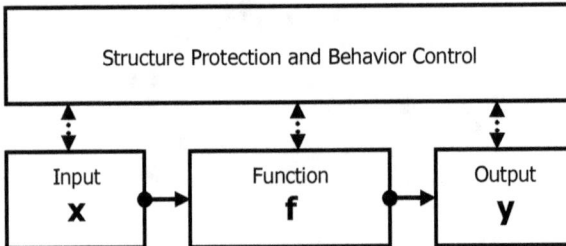

```
┌──────────────────────────────────────────────┐
│      Structure Protection and Behavior Control │
└──────────────────────────────────────────────┘
     ▲┆           ▲┆            ▲┆
     ┆▼           ┆▼            ┆▼
┌──────────┐ ┌──────────┐ ┌──────────┐
│  Input   │ │ Function │ │  Output  │
│    x     │ │    f     │ │    y     │
└──────────┘ └──────────┘ └──────────┘
```

$$y=f(x)$$

Systems with two sided\ closed protection and control mechanism
In this case (like in living beings and for example humans), mechanism has internal aspect and with feedback and flexibility. So, system can to adapt to the conditions.

Complete System Diagrams

Diagrams that can display the structure and structural order, behavior and behavioral rule and the mechanism of protecting the structural order and controlling the behavioral rule from disciplinal views, can be considered as complete system diagrams. These diagrams are a combination of the block (structural) and flow (behavioral) diagrams and also the feed-forward\ feed-back loops of information (disciplinal) diagrams. System diagrams usually include behavioral processes in the form of "Input – Function\ Process – Output" with "matter" and "energy", accompanied with the "feed-forward\ feed-back of information".

```
┌─────────────────────────────────────────────────┐
│              Protection and Control               │
│                 Orders and Rules                  │
└─────────────────────────────────────────────────┘
                         ⬍
┌─────────────────────────────────────────────────┐
│              Protection and Control               │
│                    Mechanism                      │
└─────────────────────────────────────────────────┘
      ⬍                  ⬍                  ⬍
┌──────────┐   ┌──────────────────┐   ┌──────────┐
│          │   │   Transforming   │   │          │
│  Input   │──▶│ Inputs to Outputs│──▶│  Output  │
│          │   │                  │   │          │
└──────────┘   └──────────────────┘   └──────────┘
                   ⬍         ⬆
              ┌──────────────────────┐
              │    Input - Output    │
              │        Store         │
              │(Matter, Energy and   │
              │   Information)       │
              └──────────────────────┘
```

Matter and Energy Flow ────▶

Information Flow ·······▶

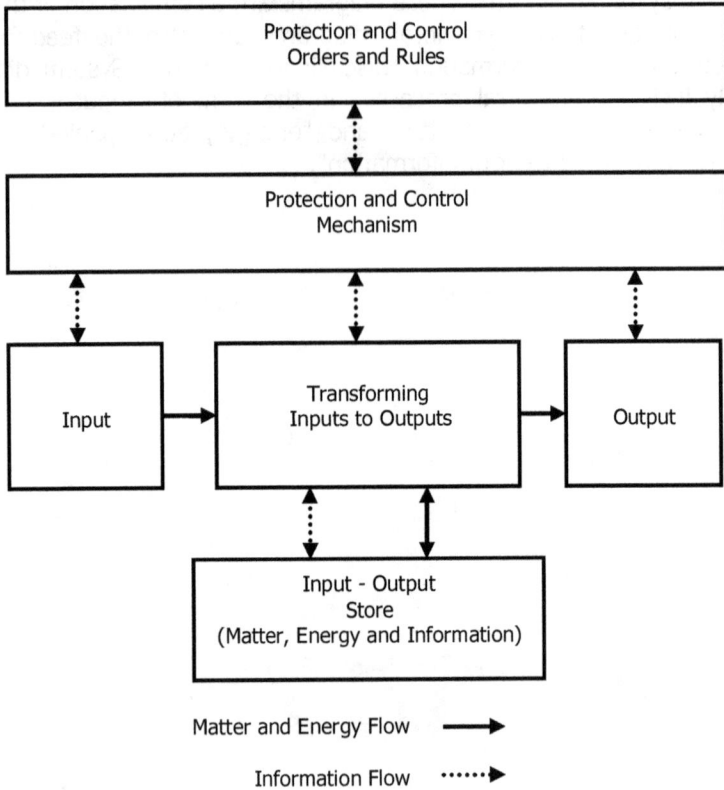

In this diagram structure, behavior and discipline of a system in a general concept is displayed. Most of the systems (like a factory) have such a total general configuration. First, the inputs to the system including the matter, energy and information are stored in their special places (like the store of raw materials). Then gradually by bringing the stored inputs to the area of operations (like the assembly line), inputs transformed to the outputs and stored in their special places (like the store of products). Finally, outputs are exited from the system. In all of these steps, the protection and control mechanism (like the organization and management of the factory), does the necessary protection and control operations by the use of the related orders and rules.

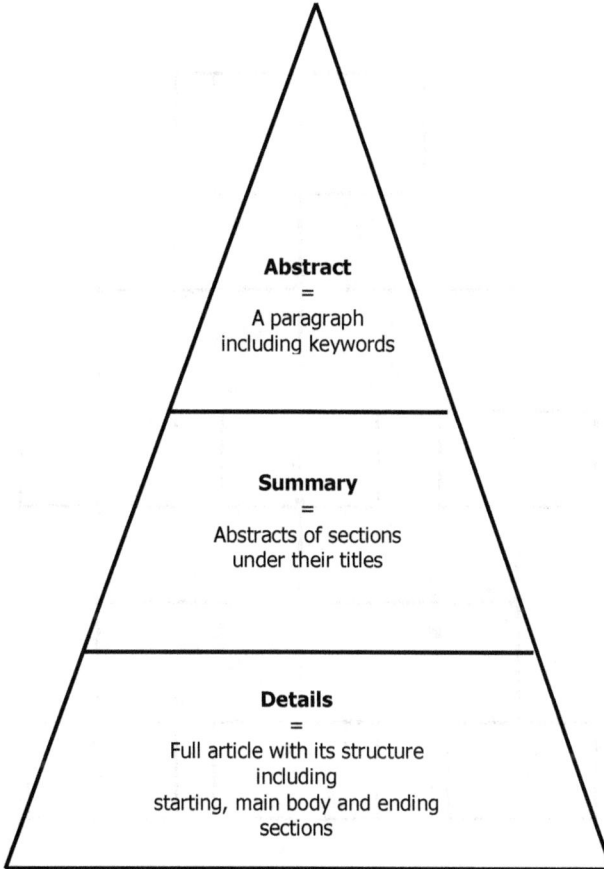

Abstract
=
A paragraph
including keywords

Summary
=
Abstracts of sections
under their titles

Details
=
Full article with its structure
including
starting, main body and ending
sections

In this diagram, different levels of describing the contents of an article is shown in "details", "summary" and "abstract" levels. This leveling also can be used in displaying the systems.

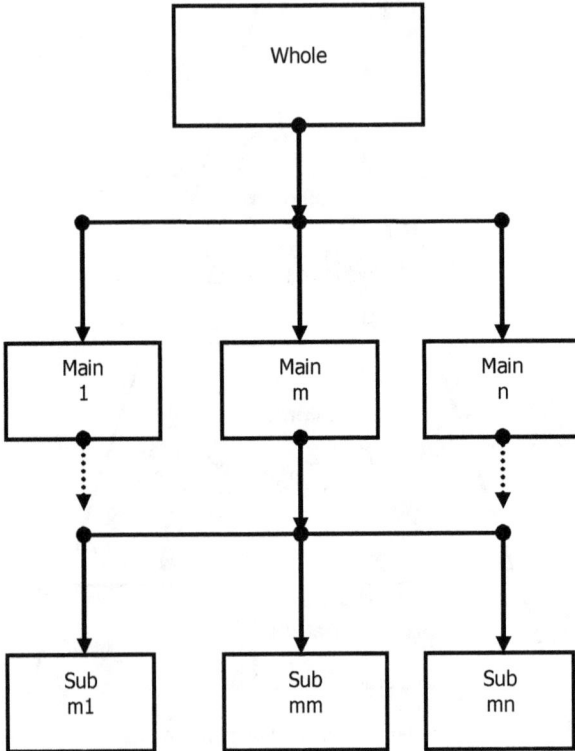

In this diagram, a system with sections in "whole", "main" and "sub" levels is displayed.

Glossary

This glossary as a reference for all of the words with special meaning in this book includes the definitions of the words delimited with "" in the text. Antonym of words (if any in this glossary), are noted in the beginning of definition of the word by "Antonym:...". For this glossary, many dictionaries have been used. Full list of the dictionaries is in the references list.

A

Abstract
1-A short text in few sentences including keywords pointing to the content of a long text.
2-Antonym: Concrete. Conceptual, understandable subjectively or logically.

Abstract System
Antonym: Concrete System. A conceptual or subjectively or logically understandable system.

Abstraction
A process in that some thing is disembodied from its numerous partial aspects and limited to a few number of its total aspects.

Action
Antonym: Reaction. Any thing that can be done.

Active Control
Antonym: Passive Control. Actuated or performed control.

Activity
Performable work or a condition so that behaviors can be done.

Activity Diagram
Diagram displaying the flow of actions during the behavior of a system.

Actual\ Real
Antonym: Imaginary\ Virtual. A really existed thing that is not imaginary and subjective.

Actuality
Antonym: Potentiality. Activated power, capacities or capabilities.

Adaptability
A process or condition in that a system in order to continue to exist and interact with the environment, coincides itself with the conditions in the environment.

Adaptive System
A system with the ability of compatibility with conditions in interacting with the environment.

Adjective

Descriptor of a noun as an object.

Adverb
Descriptor of a verb as an event or an adjective or another adverb.

Age
The length of time of existing the structure or behavior in a system.

Aggregation
A sum consisted of parts not necessarily related with each other.

Analog
Antonym: Digital. A continuous quantity or quality displayable by comparing the cases with each other, like displaying the time by comparing the places of the clock arms with the numbers on the clock screen.

Analysis
Antonym: Synthesis. The process of breaking the structure, behavior and discipline of a system into the elements and orders and also steps and rules and studying them separately, if necessary.

a part from a whole
Pointing to a portion of a totality.

Arrangement
Being besides each other with a specified order.

Atom\ Atomic
Pointing to an indecomposable object or no necessity to consider it decomposable in a specified level. The smallest possible part of a whole.

Attribute
A property or "how is" of some thing.

Automatic
Antonym: Manual. The ability of behaving in a system by using an energy source with no human being involvement.

B

Background Action
Antonym: Foreground Action. Work that is monitored indirectly or in the "back" of a state in behavior.

Balance
The equal state of all the involved masses in all sides or a condition including stability from structural view in a system.

Behavior

Using structure in order to achieve goals in a system.

Behavior Control
Directing operations by comparing it with predicted cases in its related rule.

Behavior Diagram
Diagram displaying a system functioning.

Behavior Model
Model displaying the pattern or the flow of functioning including the motion and variation in a system.

Behavioral Pattern
A specific method of doing operations in order to achieve the goals in a system while controlling the related rules.

Behavioral Rule
Methods or principals of doing operations during system functioning.

Behaviorism
Antonym: Structuralism. Giving priority to behavior and believing that the behavior is the main cause of the problems in systems.

Big Picture
Antonym: Small Picture. A view of a system with its environment.

Bilateral Symmetry
A property or structural arrangement in that the structure can be divided into two similar halves by a plane passing from the middle of the structure, like in human or animal body.

Black Box
Antonym: White Box. Pointing to the ambiguity or complexity of the subject, structure, behavior or discipline in a system.

Block Diagram
Diagram displaying the structure and behavior of a system by boxes representing the elements and arrows representing the flow of behavior.

Bottom-up Synthesis
Antonym: Top-down Analysis. The process of achieving to the whole by relating the parts to each other.

Box Diagram
Diagram displaying the basic behavioral constructs including sequence, selection and repetition in the form of boxes one after another or inside each other, also named by the name of its inventors as Nassi–Shneiderman diagram (NSD).

Brain Storming

The process of gathering and documenting all the information come to mind about some thing, free of any arrangement or organization and then trying to arrange and organize it.

C

Cases and Otherwise
A situation with the possibility of selecting one of two or more selectable cases or the otherwise case.

Cause
Antonym: Effect. Some thing as effecter of some other thing as effect.

Cause and Effect
A principle based on that any reaction is a result of an action.

Cause and Effect Diagram
Diagram displaying cause and effect process.

Cause and Effect Relation
Dependency between two things in that the second thing (effect) is existed because of the first thing (cause).

Certainty
The equal state of all the effective doubts. Stability from disciplinal view with the agency of information.

Change
The process of moving from a state to another state.

Change Rate
The degree or scale of transformation in a system.

Chaotic Change
A change without a rule or caused by an unordinary event.

Circular Cause and Effect Diagram
Antonym: Linear Cause and Effect Diagram. Diagram displaying the relation between cause and effect and also the changes resulted by effect on cause.

Circular\ Periodic Life Cycle
Antonym: Linear Life Cycle. Chain consisting of renewable steps in the life time of a system in that each step is an evolved form of the previous steps. In the case of plants, if the spring is as beginning and the fall is as ending of a step of life, then circular\ periodic life cycle is like what is seen in the nature.

Class
A set of objects or events with specific common properties under a certain title.

Closed
Antonym: Open. Absence of interaction with environment by the system or presence of control in disciplinal or cause and effect mechanisms.

Closed System
Antonym: Open System. System without any interaction with its environment in the form of matter, energy and information interchange.

Combination
Merging individually distinct things with each other.

Complex
Antonym: Simple. A situation not explainable easily.

Complex Relation
Antonym: Simple Relation. Dependency (usually non-physical) among the elements of a system not explainable easily.

Component
An important and effective part of a whole or an element of a system, replaceable with other similar one, when fails from structural or behavioral view.

Composition
Mixing things in order to have a whole mixture.

Computer Simulation
The process of making the same by imitating from real cases with the aid of computer and related software.

Concept
Meaning or a thing that can be specified and understood.

Concrete
Antonym: Abstract. Monolith. A hard and physical or objective and sensible thing with external existence.

Concrete System
Antonym: Abstract System. A monolith system. A system that exists physically or really.

Concreteness
Strength while being coordinated and compatible.

Confidence
See Certainty.

Constancy
Antonym: Variation. Situation or state with no change by time or being the same as before in different spaces and times.

Constant Structure
Antonym: Variable Structure. Structure with fixed elements and relations during behavior.

Content
Antonym: Form. Internal aspects of some thing or the things inside another thing with no direct external appearance.

Context
Environment or the conditions as the bed for some thing.

Continuous
Antonym: Discrete. A situation with no breaking. Structural elements or parts and behavioral states or steps that can not be identified clearly one by one.

Continuous Behavior
Antonym: Discrete Behavior. A behavior with no possibility of determining its states or steps certainly and one by one.

Continuous Structure
Antonym: Discrete Structure. A monolith structure with no certain borders between its parts or elements.

Control
Comparing the real cases with the predicted or desired ones from a disciplinal view in systems.

Core Processes
Main processes as backbone of behavior in systems.

Correlation
Dependency between things so that becomes stronger by time.

Crisis
Inappropriate condition or state in the structure and especially in the behavior and discipline of a system.

Current State
The present situation or step in the behavior of a system.

Cybernetics
The science of control and communication in animal and machine. Cybernetics added the information as an important factor from disciplinal aspects in systems, in order to protect the structures and to control the behaviors, besides the matter for building the structures and energy for motivating the behaviors.

Cyclic Change
Repetitive and some how regulated variations in different times.

D

Data
Facts or raw information as input for processing to produce desired outputs in a system.

Decision
The process of checking and selecting an option among a set of available options based on the current situation or condition.

Definition
A description about some thing including "What is?" "How is?" and "Why is?" from systems view.

Dependency
Relation between two things in that some of the properties or conditions in the first thing becomes related to the second thing.

Design
Activity based on certain principles, methods and tools leading to a technical description of the structure, behavior and discipline or the structure protection and behavior control mechanism in a system.

Destination
Antonym: Origination. End point in a path.

Destruction
Antonym: Growth. A natural process in that the structure, behavior or discipline of systems, from quantitative or qualitative views, gradually weakens and goes to be lesser than before.

Digital
Antonym: Analog. Numerically displayable quantity or quality like displaying the time in an electronic clock by numbers.

Discipline
Order and rule as the third aspect or "discipline" with the agency of "information" and based on the other two aspects including "structure" with the agency of "matter" and "behavior" with the agency of "energy" in systems.

Discipline\ Protection and Control Model
A model displaying the discipline or the mechanism of protecting the structural order and controlling the behavioral rule in a system.

Discrete
Antonym: Continuous. A situation with breaking and no continuance. Clearly identifiable structural elements or parts and behavioral states or steps.

Discrete Behavior
Antonym: Continuous Behavior. Behavior with certain and distinct steps or states.

Discrete Structure
Antonym: Continuous Structure. Multi part structure with certain and distinct parts or elements.

Diversity
Variety resulted from change.

Divide and Conquer
Proverb pointing to the weakness of parts against the strength of whole and in consequence the simplicity of dealing with parts against the complexity of dealing with whole and the preference of reductionism against holism.

Domain of Definition
The range of existence of structure or the realm of behavior of a system.

Dynamic
Antonym: Static. No silence and constancy as the property of systems with behavior including variation and motion.

Dynamic System
Antonym: Static System. A system with behavior based on its structure.

Dynamic View
Antonym: Static View. A view in the case of systems from behavioral aspect or variation and motion in it.

E

Effect
Antonym: Cause. Some thing resulted from some other thing.

Effectiveness
Success in fulfilling or achieving the goals.

Efficiency
Desired utility of being "enough" or "sufficient" especially in the case of structure or its order and behavior or its rule.

Element
Indecomposable structural unit. A part from a whole or a structural unit in a system without considering its decomposability.

Elements Level
Separated to constituent parts and relations among.

Embedded\ Built-in\ Natural Control

Internal control mechanism in a system originally existed or generated later.

Emergent Property
A certain property appeared in a certain level of organization in an object.

Energy
The ability of doing work. One of the three main resources (with matter and information) required for change and behavior in systems.

Entity
Any thing with its certain identity really existed or supposed to be existed in some way.

Entropy
A reverse scale (low values good and high values bad) of being weak or strong elements and relations from structural view, static or dynamic from behavioral view and order and rule or disorder and misrule from disciplinal view, naturally can be seen in all systems. In a natural trend, entropy increases in systems.

Environment
The world around some thing. A set of elements out of some thing.

Equality
Being the same from the desired points of attentions.

Equilibrium
The desired state of the involved forces from energy view including some kind of stability from behavioral view in systems.

Error
Situation implying the incorrectness or difficulty in the structure, behavior or discipline.

Evaluation
Determining or assessing of quantities or qualities in some thing by certain methods and standards to determine whether certain properties or conditions are met or not.

Event
Occurrence of action or behavior in a system.

Evolution
Antonym: Extinction. Gradually progressive development and advancement with improving in the abilities or replacement of the old with the new.

Exclusion
Antonym: Inclusion. The process of separating some thing from some other thing in order to study it, while considering its relations, role and importance.

Explicit
Antonym: Implicit. Clearly identified and without ambiguity.

Exponential Growth
Antonym: Linear Growth. Growth with increasing speed of changes.

Exterior
Antonym: Interior. Appearance of an object or system.

External Control
Antonym: Internal Control. A control mechanism in the outside of a process or system.

External Event
Antonym: Internal Event. Occurrence of an action initiated out of a system.

Extinction
Antonym: Evolution. Losing the abilities and going towards inexistence in systems.

F

Factory Settings
Rarely changing adjustments on structure and behavior, especially on discipline of technological systems set as default at the time of production.

Failure
Difficulty in the behavior and in consequence not achieving the goals in a system.

Feed-back
Antonym: Feed-forward. Returning some of the outputs again as input for disciplinal purposes in a process or system.

Feed-back Control
Antonym: Feed-forward Control. Control mechanism by returning some of the outputs again as input to the process or system for modifying the function according to the real conditions in order to achieve the goals.

Feed-back Loop
Mechanism for returning output again as input in the form of a closed loop in order to control the function.

Feed-back of Information
Mechanism of returning the data related to the current step, state or condition of a process or system as input to it for disciplinal purposes.

Feed-forward
Antonym: Feed-back. Special inputs to a process or system for disciplinal purposes.

Feed-forward Control

Antonym: Feed-back Control. Control mechanism by special inputs to a process or system for adjusting it.

Field
An area in general affected by an effecter including matter or mass, energy or force and information or the order and rule based on it.

Finite States Machine\Model = FSM
A machine or model like a home washing machine, with a behavior in the form of certain and limited states.

Fish Bone Diagram
Diagram displaying cause and effect in the form of a "fish bone" also called in the name of its inventor, "Ishikawa diagram".

Flexibility
Adaptability with the changes in structure, behavior and discipline or environment.

Flow Chart
Diagram displaying the flow of operations in the behavior of a system by special symbols.

Flow Diagram
General title of diagrams displaying the flow of behavior in systems.

Flow Rate
Amount of some thing going from a source to a destination in a unit of time.

Force
Cause of change, including change in the place, shape or speed of some thing.

Foreground Action
Antonym: Background Action. A directly monitored work in a state or step of behavior.

Form
Antonym: Content. External aspects and the appearance of some thing.

Function
1-Set of operations on inputs to produce outputs as the total behavior of a system.
2-A quantity its value depends on some other quantities as variables.

Function of Functions
A mathematical relation displaying the dependency of a quantity to some variables that are also depended on some other variables. Action resulted from other actions.

Fundamental Theory and Practice

A view and act with a wide coverage and applicable in the case of any thing. A base for research and experiment in any field.

G

General Systems Theory = GST
A theory presented by the Austrian biologist Ludwig Von Bertalanffy (1901 – 1972) in 1945. Based on this theory, approximately all the things in the world around us, are as a "system".

Generalization
Antonym: Specialization. Centralizing on total general aspects in things.

Goal
A certain preferred step or state in order to achieve in a system. Goals usually are short-term and fulfilling the goals is usually for fulfilling the mid-term purposes and long-term ideals in the behavior of systems.

Goal\ Purpose\ Ideal seeking System
A system that wants to achieve certain goals in order to fulfill certain purposes so that to get closer to its ideals that usually are not achievable in the real world. This process continues with new goals, purposes and ideals, evolutionary and progressively in all over the life time of the system.

Growth
Antonym: Destruction. A natural process in that the structure, behavior or discipline of systems, from quantitative or qualitative views gradually strengthens and evolves.

H

Hard
Antonym: Soft. Property of objects or systems with concrete or physical sensible nature.

Hard System
Antonym: Soft System. Physical sensible systems like technological devices.

Hardness
Antonym: Softness. Inflexibility due to the physical nature.

Hardware
Antonym: Software. Tools, machinery and equipment with physical nature.

Hierarchy of Systems
Leveling of systems based on the level of a system among other systems in that higher level systems include the lower level systems and use their facilities in achieving the goals.

Holism
Antonym: Reductionism. Giving priority to the whole against the parts in dealing with the objects.

Holon
Some thing may or may not be considered as a system.

Homeostatic System
A system tending to have stability by structural balance, behavioral equilibrium and disciplinal certainty, especially like in biological processes of living beings.

Homogeneity
Being the same in all points and in all sides, especially from structural view.

I

Icon
A small picture or symbol for displaying an object or event, like the ones used in weather forecasting.

Ideal
Most desirable step or state in the behavior of a system as ultimate goals that may not be achieved in practice.

Identifiable
A property based on the ability of recognizing some thing.

Identification
The process of recognition of some thing.

Identification Key
A unique data like the social security number of some one that uniquely represents a certain thing.

Identifier
An indicator that indicates some thing.

Imaginary\ Virtual
Antonym: Real\ Actual. A thing that does not exist in the external world and needs mental representation.

Implementation
The process of executing or realization of the design in the case of a system and creating it in practice.

Implicit
Antonym: Explicit. Stated in the form of another thing and with the possibility of ambiguity.

Inclusion
Antonym: Exclusion. A process in that some thing can encompass some other things.

Independency
Antonym: Dependency. Not depending on some other thing, especially from behavioral and disciplinal control aspects.

Informatics
Combination of "information" and "automatic" as the automatically processing of information.

Information
One of the three main and basic resources next to the matter and energy, needed for discipline and certainty in systems.

Initial Settings
Adjustments on structure, behavior or discipline of a system, set on it according to the environmental conditions during its behavior or functioning.

Initial\ Start State
Antonym: Terminal\ Stop State. A situation with certain conditions at the start of behavior of a system.

Input
Antonym: Output. Matter, energy and information in general, entered to a system for producing outputs.

Input – Function – Output = IFO
Chain of the main operations as the backbone of function in processes and behavior in systems.

Input\ Output
The sides of function in process or behavior in system, pointing to what is put in and what is put out.

Inspection
Checking the case led to error or failure in system.

Instance
A thing among the many similar things in a set or class of things.

Interaction
Action with reaction related with each other.

Intercourse
Interactions including give and get between two sides.

Interdependency

Antonym: Independency. Mutual relation in the things, especially from behavioral and disciplinal control aspects.

Interface
A place that two or more related systems get together and the adjustments on systems set there.

Interior
Antonym: Exterior. Internal side of an object or system.

Internal Clock\ Timer
Internal mechanism for measuring or managing the time in the systems with behavior in time.

Internal Control
Antonym: External Control. A control mechanism inside of a process or system.

Internal Event
Antonym: External Event. Occurrence of an action initiated from inside of a system.

Invisible
Antonym: Visible. Some thing unidentifiable in ordinary ways.

Irregular
Antonym: Regular. No rule in acting.

Isolated and Separated
Antonym: Organized and Related. Unrelated parts or elements without wholeness as a system.

Iterative and Progressive
Renewal of a process in order to produce more desirable output.

K

Keyword
A word pointing the content of a text.

L

Layered Structure
Structure organized in different levels.

Learning
The process of knowledge acquisition with the aid of memory by study, experience or interacting with the environment.

Life Cycle

The chain of creation of structure and presentation of behavior under the related discipline in order to achieve the goals in systems once for ever or in an iterative form.

Life Time
The whole duration of existence of structure and presentation of behavior in systems.

Linear Cause and Effect Diagram
Antonym: Circular Cause and Effect Diagram. A diagram in that the cause is related with its effect by an arrow and the change on cause due to effect is ignored.

Linear Chain of Sequential Systems
A set of systems one after another so that each system can use the output of other system as input. In this chain of systems, raw initial inputs at the beginning, transformed to the processed final outputs at the ending of chain.

Linear Growth
Antonym: Exponential Growth. Fixed speed growth.

Linear\ Start – Stop Life Cycle
Antonym: Circular\ Periodic Life Cycle. Life with distinct begin and end consisting of "birth", "living" and "death" in systems.

Liquidity
Antonym: Rigidity. Flexibility of structure against the changes in behavior.

Logical
Antonym: Physical. Conceptual, subjective or understandable.

Logical Model
Antonym: Physical Model. Model based on the logically understandable elements.

Logical Motor\ Engine
Antonym: Physical Motor\ Engine. Driver with soft or understandable aspect like the internet search engines.

Long Term Behavior Pattern
A long time behavioral method some how institutionalized in a system.

M

Macro
Antonym: Micro. Some thing in a scale greater than the regular common scales.

Main System
Antonym: Sub-system. A big and macro system with small and micro systems as the sub-systems in it.

Maintenance
Antonym: Repair. The process of keeping the "ready to work" state or continuing the behavior in systems in order to prevent the probable errors or failures.

Management
The process of governing the objects, events and their orders and rules.

Manual
Antonym: Automatic. Work or behavior with the aid of human being.

Mass
The amount of matter in an object.

Master Data = Fixed Basic Information
Antonym: Transaction Data = Variable Current Information. Fixed data in the whole or a long time of the life of a system identifying its identity or the structure and behavior of it.

Matter
Maker of the sensible body of any thing or a physical substance in general or any arguable subject.

Measure of Performance
A macro scale as the resultant of other micro scales about the behavioral or functional aspects of a system.

Measurement
Determining the amount or quantity of some thing by using appropriate tool and unit.

Mechanism
Supplies and methods of doing work in the form of a structure with a machine-like behavior.

Member
Someone from a set, a part from a whole, an element from a system or a section of a structure.

Memory
Facilities in the form of an organ in animals or a section in machines as a place and mechanism for recording, keeping and accessing information during behavior.

Meta Data
Data that consists other data.

Meta Element
Element that consists other elements.

Meta System

System that consists other systems.

Method
Specific way of doing work or behaving.

Methodology
A set of methods and tools applicable in a certain field of study or work.

Micro
Antonym: Macro. Some thing in a scale smaller than the regular common scales.

Micro\ Macro Element
Element in a scale smaller\ greater than the regular common scales.

Micro\ Macro System
System in a scale smaller\ greater than the regular common scales.

Mid State
General title of the behavioral situations between the two special situations of start and stop of behavior.

Mobility
The ability of change and diversity in the space or location.

Model
Objective or subjective abstraction of the important generalities of some thing without emphasizing on details.

Modeling
The process of making model as the objective or subjective abstraction of some important aspects of a thing without emphasizing on details.

Monitoring
Evaluating the quantities or qualities of input, function and output of a process or system in order to improve or make decision about it.

Multi State
Antonym: Single State. Situations with different conditions in different times in the behavior of a system.

Multiple Relations
Antonym: Single relation. Numerous relations among the elements in a system.

Multiplicity
Antonym: Unity. Diversity in the cases.

N

Negative\ Subtractive

Antonym: Positive\ Additive. Decreasing effect in related changes with increase in one, decrease in the other one.

Nested Sets
Groups of things consisted of smaller groups of things in lower levels.

Normal
Some kind of un-written or natural standard that is obvious or appears as "must be".

Noun
A word in the language to point to an objective or subjective thing.

Numeric Value
Any assumed quantity for some thing in the form of figures.

O

Object
1-Any objective or subjective thing, sensed or understood in some way.
2- Antonym: Subject. Some one or some thing that a verb or behavior is done on it.

Object Model
A model that presents an object as it is or imagined. Model of the elements of a system.

Objective
Antonym: Subjective. A real and sensible thing without the need to be imagined.

On\ Off
Two contradictory states implying the state of behaving or not behaving in systems.

Open
Antonym: Closed. Presence of interaction with environment by the system or absence of control in disciplinal or cause and effect mechanisms.

Open System
Antonym: Closed System. System with interactions in the form of matter, energy and information with the environment.

Operations
Set of actions for achieving certain goals.

Order
Arrangement or spatial fixation of the elements that get-together and relate in order to create a whole.

Organicism
Giving priority to the information and discipline in systems and believing that the root of most of the problems and difficulties in systems are due to lack of true discipline or the weakness of organization and management.

Organic System
A system like living beings and the use of information in it like the use of matter and energy.

Organism
Thing, system or organization like what is in the living beings from structural, behavioral and disciplinal views.

Organization
A set of related elements around certain goals and behaving as a living being by information interchange.

Organizational – Managerial Approach
Dealing with the objects by considering them as living beings or with structural organization and behavioral management like in human societies.

Organizational View
A view in the case of systems with considering the system of interest with its structure, behavior and discipline, like a living being.

Organized and Related
Antonym: Isolated and Separated. Related elements with wholeness as a system.

Organizing
The process of relating the elements around certain goals and behaving as a living being, by information interchange.

Origination
Antonym: Destination. Start point in a path.

Oscillation Amplitude
Changes between up and down in a wave form change or behavior.

Output
Antonym: Input. Processed matter, energy and information as the product of function or behavior of a process or system.

P

Part
Antonym: Whole. Section or portion of a whole.

Passive Control
Antonym: Active Control. Potential and not actually performed control.

Pause State
Temporarily suspend of system behavior for some reasons.

Performance
Doing a work or activity according to a rule or a plan.

Permanency
Antonym: Temporality. Continued state or situation in all over a considered time length.

Permanent Relation
Antonym: Temporary Relation. Dependency between two or more elements continuing all over the considered time length.

Permutation
Things besides each other with considering the order of the things.

Perspective
A sight or point of seeing some thing.

Phenomenon
Some thing appeared or occurred amazingly.

Physical
Antonym: Logical. Real, objective or sensible.

Physical - Logical Model
Model based on the physically sensible and logically understandable elements.

Physical Model
Antonym: Logical Model. Model based on the physically sensible elements.

Physical Motor\ Engine
Antonym: Logical Motor\ Engine. Driver with hard or sensible aspect like the ordinary car engine.

Picture Diagram
Diagram by using symbolic pictures and arrows.

Plan
Technical explanation and presentation of structure, behavior, discipline and related protection and control mechanisms in the case of a system by using certain principles, methods and tools.

Positive\ Additive
Antonym: Negative\ Subtractive. Increasing effect in related changes with increase in one, increase in the other one.

Post-condition
Antonym: Pre-condition. Condition governing or must be reached after moving from one to another state or an event occurrence.

Potentiality
Antonym: Actuality. Not activated power, capacities or capabilities.

Pre-condition
Antonym: Post-condition. Condition governing or must be reached before moving from one to another state or an event occurrence.

Prefect Symmetry
Structural arrangement with equality and similarity in all sides, like sphere and around the center of it.

Primary Fundamental Order and Rule
Antonym: Secondary Defined Order and Rule. Basic order and rule due to the nature of the things.

Procedure
Stepwise method for doing a certain work.

Process
Chain of the actions including getting inputs, functioning on inputs to produce outputs and giving outputs.

Process Flow Diagram
Diagram displaying the flow of input, function and output in a process.

Processed Data
Produced information useable in decision makings.

Processing
Transforming primary raw materials to secondary produced products in a process.

Processing Input to Produce Output
Acting on what is gotten to produce what is given.

Productivity
The ability of behaving with desired results more than the norm.

Profitability
The ability of behaving with the benefits of the outputs more than the benefits of inputs.

Property
Attribute of some thing for describing it.

Protection and Control Diagram

Diagram displaying the mechanism of structural order protection and behavioral rule control in a system.

Protection and Control Mechanism
Schemes and methods that protect or maintain the structural order and control or direct the behavioral rule in a system.

Purpose
Reason for the existence of some thing.

Q

Quality
Antonym: Quantity. The "how value" of some thing by mentioning a kind of value for it.

Quantity
Antonym: Quality. The "what amount" of some thing by mentioning a kind of amount for it.

R

Radial Symmetry
A structural property or arrangement with equality and similarity around an axis passing from the center of the structure.

Range
The limits of changes between minimum and maximum in the case of a measurable quantity.

Reaction
Antonym: Action. An event or work that may be occurred or done against other event or work.

Real\ Actual
Antonym: Imaginary\ Virtual. A thing that exists in the external world and does not need to imagination in mind.

Realization
Making operational an idea or system according to the design based on the studies about it.

Reductionism\ Elementalism
Antonym: Holism. Giving priority to the parts against the whole in dealing with the objects.

Regular
Antonym: Irregular. Having rule in acting.

Relation
Dependency of things with each other.

Repair
Antonym: Maintenance. Making a destructed thing well so that to operate as before.

Repetition
Redoing a set of actions for some purposes.

Replace ability
The ability of substitution of some thing in place of another thing.

Request
Antonym: Response. Asking for some thing.

Response
Antonym: Request. Answer against an asking.

Rigidity
Antonym: Liquidity. Inflexibility of structure against the changes in behavior.

Role
Position of an element in doing the tasks in a system.

Routine
A settled down method for doing a work.

Rule
Principles of presenting the behavior according to the steps in time.

S

Secondary Defined Order and Rule
Antonym: Primary Fundamental Order and Rule. Created order and rule during the existence of some thing.

Selection
Choosing one of the two or more options based on the conditions.

Self Organizing & Self Managing Systems
Systems with the ability of ordering the structure and protecting it and ruling the behavior and controlling it.

Semantic
Subject related with the concept of words in sentences in the language.

Semiotic
Subject related with the symbols and its interpretations and applications.

Sequence
Set of actions done one after another.

Set
Collection of things under a name or title.

Set of Elements
Collection of parts.

Set of Functions
Collection of related mathematical equations.

Set of Sets
Collections including other sub-collections.

Set Theory
Theory related with the collection of objects and its properties as the base of modern mathematics.

Similarity\ Resemblance
Being alike from specific aspects.

Simple
Antonym: Complex. Easy situation.

Simple Relation
Antonym: Complex Relation. Easy dependency among the elements in a system that usually are physical dependencies.

Simulation
Process of displaying the structure, behavior or discipline of a system by imitating it from the system.

Single Relation
Antonym: Multiple Relations. Individuality in dependency between the elements in a system.

Single State
Antonym: Multi State. A system always in one situation.

Small Picture
Antonym: Big Picture. A view of a system without its environment.

Soft
Antonym: Hard. Property of objects or systems with abstract or logically understandable nature.

Soft System

Antonym: Hard System. Logically understandable systems like virtual machines or the societies of living beings.

Softness
Antonym: Hardness. Flexibility due to the logical nature.

Software
Antonym: Hardware. Methods, procedures and programs with logical nature.

Space
Continuum of infinite number of points in all sides in that the objects have the possibility of appearance.

Specialization
Antonym: Generalization. Centralizing on partial special aspects in things.

Stability
Desired constancy in systems including structural balance, behavioral equilibrium and disciplinal certainty.

Stable State
Situation or state with desired conditions in systems.

Stand Still
Always remained situation.

Stand-by State
Temporarily stopped behavior with the possibility to resume it in the behavior of technological systems.

Start\ Stop
Two important points in the behavior of systems without any supposed action or reaction in the form of behavior before start and after stop.

State
Situation with specific conditions in a specific moment of time in the behavior of a system.

State Entry Action
An action that can or must be done when entering to a new situation in the behavior.

State Exit Action
An action that can or must be done when exiting a situation in the behavior.

State Maintaining System
A system that always can keep its existing situation.

State Stop Action

An action that can or must be done when staying in a situation in the behavior.

State Transition
Going from existing situation to another situation in the behavior of a system.

State Transition Diagram = STD
Diagram displaying all the possible behavioral situations of a system with the required actions for going from one situation to another.

States Chart
Diagram displaying all the possible behavioral situations of a system.

Static
Antonym: Dynamic. Silent, remained constant or with no change as the property of a system with structural aspect as its main aspect.

Static System
Antonym: Dynamic System. System summarized only in its structural aspect.

Static View
Antonym: Dynamic View. A view about the systems by that the system is important only from structural aspects including silence and constancy.

Steady Change
Permanent process of change possibly due to the nature of the related thing.

Steady State System
System always in a permanent constant situation and without the known processes like the growth and evolution or destruction and extinction in it.

Stock and Flow
Mechanism ruling most of the processes in the world around and including "flows" from sources to destinations as "stocks".

Stock and Flow Diagram
Diagram displaying the flows from stocks to other stocks until consuming.

Storage of Inputs
Store of the entered things in order to function on it to produce the desired exiting things.

Storage of Outputs
Store of the results in order to give it out.

Strength
Antonym: Weakness. Power in structure, behavior and discipline and ability in function and resistance against unusual changes.

Strong Relation

Antonym: Weak Relation. Powerful dependency among the elements and resistant even against unusual changes.

Structural Element
Units making the structure.

Structural Order
Arrangement of elements of a system to create its structure and maintain its structural balance.

Structural Property
Property of the combined parts to create the whole as a system.

Structural Uniformity
Similarity of constituent parts and relations of the whole in all sides.

Structural Wholeness
Totality due to the relations among the elements as a certain thing.

Structuralism
Antonym: Behaviorism. Giving priority to structure and believing that structure is the main cause of the problems in systems.

Structure
1-Spatial arrangement of the parts or elements and relations among as the base for existence of an object or system.
2-Buiding or any thing with specific architecture and made of specific substances.

Structure Diagram
Diagram displaying the spatial arrangement of the constituent parts or elements and relations of an object or system.

Structure Model
Model displaying the features including the silence and constant aspects of some thing.

Structure Protection
Keeping the arrangement and order of the parts or elements of an object or system in order to keep the wholeness and balance in it and preventing it from collapsing.

Structured
Any thing that its constituent parts or elements arranged or ordered in a specific form.

Study
Reading and thinking about different aspects of some thing by referring to the existed references and documents about it.

Sub-domain
A part from the structural scope, behavioral realm or disciplinal governance.

Subject
1- Antonym: Object. Noun or noun implying clause including the name of the actor of an action.
2-Title or some thing that can be discussed or studied.

Subjective
Antonym: Objective. A conceptual thing that needs to be imagined mentally.

Sub-system
Antonym: Main System. A system inside a bigger system.

Summary
A text based on the structure of an article that presents the content of it as the abstracts of its sections in short.

Super Element
Element consisted of other elements.

Super System
System consisted of other systems.

Symbol
A sign or thing representing some other thing.

Symmetry
Structural arrangement with some kind of equality and similarity in two opposite (right and left or up and down) or more sides.

Syntax
Grammar of the language or the rule for combining the words to form sentence in language.

Synthesis
Antonym: Analysis. Relating the parts in order to form the whole consisting of the parts.

System
A set of related elements as a whole with "structure" and "structural order", "behavior" and "behavioral rule" and "discipline" including the mechanism of "protecting structural order" and "controlling behavioral rule".

System Behavior
Action and reaction, cause and effect or the process of input, function and output in systems.

System Boundary

The border that surrounds or encompasses the elements of a system and separates it from the environment.

System Discipline
The order governing the structure, the rule governing the behavior and the mechanism of protecting and controlling it in systems.

System Domain
Structural scope, behavioral realm or disciplinal governance of a system.

System Dynamics
Topic related with the behavior or changes in systems.

System Environment
The bed of system and the source of system inputs and destination of system outputs.

System Level
Seeing the system in its wholeness.

System Life
Existence of structure and presentation of behavior while protecting the structure and controlling the behavior in a system.

System Model
Model representing the whole of a system consisting of structure, behavior and discipline.

System of Interest
System as the case or the subject of study of a researcher.

System Perspective
View of seeing a system.

System Statement
An explanation of some thing as a system.

System States Model
Model presenting the various situations of a system during its behavior.

System Structure
Spatial arrangement of the elements and relations as the base of existence of a system.

System Total Behavior
General function of a system as the emergent and pervasive function in all conditions.

System User\ Actor

Agent that acts or reacts with a system in the form of system behavior or function.

Systematic\ Rule Based Approach
Method of dealing with the events by considering the governing rule in the behavior resulted from the event.

Systemic\ Order Based Approach
Method of dealing with the objects by considering the governing order in the structure resulted from the object.

Systemism
Believing that any entity is considerable as a system and explainable by the system concepts.

Systems Analysis and Synthesis
The process of reaching to the parts from breaking the whole and reaching to the whole by relating the parts.

Systems Approach
Method of dealing with the objects and related events as systems.

Systems Engineering
A branch of engineering with the subject of systems.

Systems Methodology
Set of methods and tools for using in systems approach or thinking about the objects and related events.

Systems Re-engineering
Reviewing the existing systems and re-studying, re-designing and re-creating them based on the new scientific and technological achievements in the related fields.

Systems Thinking
Method of thinking based on the system concept in dealing with the objects and related events.

T

Temporality
Antonym: Permanency. State or situation not continuing in all over the considered time length.

Temporary Relation
Antonym: Permanent Relation. Dependency between two or more elements not continuing in all over the considered time length.

Terminal\ Stop State
Antonym: Initial\ Start State. A situation with certain conditions at the end of behavior of a system.

Theme
A main or central subject with other subjects around it.

Theory for Everything
General and universal concepts of system, systems thinking and General Systems Theory (GST) that can be used approximately in the case of every thing.

Thermostat
Mechanism that automatically activates or deactivates a device based on a certain degree in order to keep the temperature in the same level.

Thing
An object, event or subject in general when can not be called by a certain name or title.

Thinking Pattern
A conventional method for investigation about the objects and events in a certain format.

Throughput
Potential, capacity or the speed and intensity of converting inputs to outputs in functions.

Time
Continuum of infinite number of moments in that the events have the possibility of occurrence, continued in the past, continues at the present and will continue in the future.

Time Invariant
Antonym: Time Variant. A situation in the case of some thing that does not change by time.

Time Variant
Antonym: Time Invariant. A situation in the case of some thing that changes by time.

Top-down Analysis
Antonym: Bottom-up Synthesis. The process of reaching to the parts by breaking the whole into its parts.

Topology
A branch of mathematics in that the geometrical properties and spatial relations of an object, not changed by a continuous change in the shape or size of the object, are studied.

Total System
A system composed of other systems and acting as a main system with inputs, functions and outputs of the sub-systems inside it.

Transaction Data = Variable Current Information
Antonym: Master Data = Fixed Basic Information. Variable data about an object or its behavior in the whole or periods of life of the object.

Transformed\ Processed Resource
Output form of matter, energy or information.

Transition Rule
Method or principle of going from one state or step to another state or step in the behavior.

Trap
Unwanted situation in the behavior, so that escaping from it is difficult or impossible.

Two State
Behavior or system containing only two possible steps or situations like "on" and "off" or "yes" and "no".

U

Unity
Antonym: Multiplicity. Union of parts in order to create a whole.

Universal Set
A set (named U set) as the environment of all of the other sets.

Unusual
Antonym: Usual. Any uncommon or abnormal thing.

Use Case
Method or tool presenting the interactions with a system by its user.

Usual
Antonym: Unusual. Any common or normal thing.

Utopia
An imaginary place with the best ever possible structure, behavior and discipline in it.

V

Variable Structure
Antonym: Constant Structure. Structure with elements and relations not fixed during behavior.

Variation

Antonym: Constancy. Situation or state with change by time or being different in different spaces and times.

Verb
A word in the language to point to doing some thing or occurrence of an event.

View
Seeing some thing from a specific sight.

Virtual\ Imaginary
Antonym: Real\ Actual. Thing that does not exist in the external world and needs to be imagined in mind.

Visibility Rate
The amount of clearness of some aspects in a thing.

Visible
Antonym: Invisible. Some thing identifiable in ordinary ways.

Visual Modeling
The process of making model in the form of picture or diagram by using certain symbols for certain things.

W

Weak Relation
Antonym: Strong Relation. Dependency among the elements with the possibility of breaking even under usual changes.

Weakness
Antonym: Strength. Looseness in structure, behavior and discipline and disability in function and breaking under usual changes.

White Box
Antonym: Black Box. Pointing to the clearness or simplicity of the subject, structure, behavior or discipline in a system.

Whole
Antonym: Part. A thing consisted of related parts.

Wholeness
Totality due to the relations among its parts.

Y

Yes\ No
Two common states or responses indicating two contradictory states against each other as turning point in behavior.

References and further readings

Books & Articles

A

*Abramovitch, D., ..., The Outrigger: A Prehistoric Feedback Mechanism, Agilent Laboratories, http://www.labs.agilent.com/personal/Danny_Abramovitch/pubs/outrigger_hist_mat3.pdf

*Ackoff, R., ..., Thinking about the future, http://ackoffcenter.blogs.com/ackoff_center_weblog/files/ackoffstallbergtalk.pdf

*Ackoff, R., ..., Why Few Organizations Adopt Systems Thinking, ..., http://ackoffcenter.blogs.com/ackoff_center_weblog/files/Why_few_aopt_ST.pdf

*Ackoff, R., 1963, General Systems Theory and Systems Research: Contrasting Conceptions of Systems Science, General Systems Year Book, Vol. 8, pp. 117–121.

*Ackoff, R., 1971, Towards a System of Systems Concepts, Management Science, Jul. 1971, pp. 661-671, http://ackoffcenter.blogs.com/ackoff_center_weblog/files/AckoffSystemOfSystems.pdf

*Ackoff, R., 1974, The Systems Revolution, Long Range Planning, Vol. 7, No. 6, Dec. 1974, pp. 2-20, http://www.sciencedirect.com/science/article/pii/0024630174901277

*Ackoff, R., Addison, H. J., 2006, A Little Book of f-Laws: 13 common sins of management, Triarchy Press, United Kingdom, http://www.f-laws.com/pdf/A_Little_Book_of_F-LawsE.pdf

*Albin, S., 1997, Building a System Dynamics Model, MIT, http://ocw.mit.edu/courses/sloan-school-of-management/15-988-system-dynamics-self-study-fall-1998-spring-1999/readings/building.pdf

*American Academy of Child and Adolescent Psychiatry, 2009, Systems-Based Practice Overview, American Academy of Child and Adolescent Psychiatry.

*Aronson, D., 1996, Overview of Systems Thinking, Thinking Page, http://www.thinking.net/Systems_Thinking/OverviewSTarticle.pdf

*Ashby, R., 1957, An Introduction to Cybernetics, London, http://pespmc1.vub.ac.be/books/IntroCyb.pdf

*Auyang, S. Y., 2004, Concepts of System in Engineering, talk presented at Department of the History of Science, Harvard University, http://www.creatingtechnology.org/eng/system.pdf

B

*Bailey, K. D., ..., Entropy Systems Theory, Encyclopedia of Life Support Systems, http://www.eolss.net/Sample-Chapters/C02/E6-46-01-04.pdf

*Bartlett, G., 2001, Systemic Thinking, International Conference on Thinking "Breakthroughs 2001", Probsolv International, http://www.probsolv.com/systemic_thinking/Systemic%20Thinking.pdf

*Barton, J., Ryan, T., Pragmatism: Systems Thinking and System Dynamics, Monash University, Australia, http://www.systemdynamics.org/conferences/1999/PAPERS/PLEN2.PDF

*Bell, D., 2003, UML Basics Part II: The activity diagram, The Rational Edge Ezine for Rational Community, http://www.ibm.com/developerworks/rational/library/content/RationalEdge/sep03/f_umlbasics_db.pdf

*Benjamin, C., Jones, A., 2006, Systems Thinking: A Practical Application, Sustainability Institute, http://www.sustainer.org/pubs/SI06JonesBO1Final.pdf

*Bergvall, Kareborn, B., 1999, Book review of "Checkland, P. and Holwell, S., 1998, Information, Systems, and Information Systems, Chichester, John Wiley & Sons", Cybernetics & Human Knowing, 1999, Vol. 6, No. 3, pp. 91–95, http://www.imprint.co.uk/C&HK/vol6/check-review_6-4.PDF

*Bertalanffy, V., 1968, General Systems Theory: Foundations, Development, Applications, Braziller, New York, ISBN for 2003 print: 0807604534, 9780807604533, http://ebookbrowse.com/adv.php?q=Bertalanffy%2C+V.%2C+1968%2C+General+Systems+Theory%3A+Foundations%2C+Development%2C+Applications+ebook&source=1

*Binder, T., Vox, A., Belyazid, S., Haraldsson, H. and Svensson, M., ..., Developing System Dynamics Models From Casual Loop Diagrams, ...

*Blauberg, I. V., Sadovsky, V. N., Yudin, B. G., 1977, Systems Theory: Philosophical and Methodological Problems, Progress, Moscow.

*Bogdanov, A. A., 1922, The Universal Science of Organization: Essays in Tektology, Berlin and Petrograd-Moscow (in Russian), Translated and reprinted as Dudley, P., 1996, Bogdanov's tektology, Centre for Systems Studies, University of Hull, http://books.google.com/books/about/Essays_in_tektology.html

*Borshchev, A., Filippov, A., ..., From System Dynamics and Discrete Event to Practical Agent Based Modeling: Reasons, Techniques, Tools, Petersburg Technical University, Russia, http://www.systemdynamics.org/conferences/2004/SDS_2004/PAPERS/381BORSH.pdf

*Boulding, K., 1956, General Systems Theory: The Skeleton of Science, Management Science, Apr. 1956, pp. 197-208, http://web.ku.edu/~jleemgt/MGMT%20916/PDF/Boulding1956MS.pdf

*Boxer, PJ., 1994, Checkland: Soft Systems Methodology, ..., http://www.it.murdoch.edu.au/~sudweeks/b329/readings/boxer.pdf

*Bunge, M., 2000, Systemism: The alternative to individualism and holism, Journal of Socio-Economics Vol. 29 ,2000, pp. 147–157,

*Burton, M., 2003, Book review of "Midgley, G. 2003, Systemic Intervention: Philosophy, Methodology, and Practice", Journal of Community and Applied Psychology, 2003 Vol. 13 No. 4, pp. 330-333, http://users.cooptel.net/mark.burton/ReviewMidg.pdf

C
*Carter, T., 2011, An Introduction to Information Theory and Entropy, Complex Systems Summer School, Santa Fe, http://astarte.csustan.edu/~tom/SFI-CSSS/info-theory/info-lec.pdf

*Centre for Organization Development, ..., A Systems Thinking Approach, Centre for Organization Development Pty Ltd., http://www.cfod.com.au/wp-content/uploads/2011/06/Intro-to-OD.pdf

*Charlton, B. G., Andras, P., 2003, What is management and what do managers do? A systems theory account, Philosophy of Management, 2003, No. 3, pp. 1-15, http://www.hedweb.com/bgcharlton/rip-management.html

*Chatzkel, J. L., Chatzkel, B., ..., How Can Systems Thinking Enhance Quality Efforts? Conference on "Systems Thinking in Action" explores organizational learning structures, http://www.progressivepractices.com/articles/systems_thinking.pdf

*Checkland, P., 1981, Systems Thinking, Systems Practice, Wiley, ISBN for 1981 print: 0471279110, 9780471279112, http://books.google.com/books/about/Systems_thinking_systems_practice.html

*Checkland, P., 1999, Systems Thinking, Systems Practice: A 30 year retrospective, Wiley, Chichester, http://books.google.com/books/about/Systems_Thinking_Systems_Practice_Includ.html

*Checkland, P., Holwell, S., 1998, Information, Systems, and Information Systems, Chichester, John Wiley & Sons, ISBN. 0471958204, 9780471958208, http://books.google.com/books/about/Information_systems_and_information_syst.html

*Checkland, P., Scholes, J., 1990, Soft systems methodology in action, Wiley, Toronto, ISBN for 1990 print: 0471927686, 9780471927686, http://books.google.com/books/about/Soft_Systems_Methodology_in_Action.html

*Chen, Z., Reigeluth, C. M., 2010, Communication in a Leadership Team for Systemic Change in a School District, Contemporary Educational Technology, 2010, Vol. 1, No. 3, pp. 233-254, http://www.cedtech.net/articles/13/134.pdf

*Churchman, W., 1979, The Systems Approach, New York, Dell Publishing, ISBN. 0385289987, 978-0385289986, http://www.amazon.com/Systems-Approach-Charles-West-Churchman/dp/0385289987

*Clinton, K. L., Dougherty, T. A., Masimore, T. R., 2006, Systems Thinking: An Effective Tool for Problem Resolution and Change Management, Systems Research Forum, Vol. 1, No. 1, 2006, http://www.worldscientific.com/doi/abs/10.1142/S1793966606000084

E
*Einstein, A., 1941, Science and Religion, talk presented at Science, Philosophy and Religion Symposium, New York, http://www.update.uu.se/~fbendz/library/ae_scire.htm

*Ericsson, M., 2004, Activity Diagrams: What it is and How to Use, IBM, Developers Work,

http://sunset.usc.edu/classes/cs577a_2000/papers/ActivitydiagramsforRoseArchitect.pdf

F

*Fliessbach, K., Weis, S., Klaver, P., Elger, C.E., Weber, B., 2006, The effect of word concreteness on recognition memory, NeuroImage 32 ,2006, pp. 1413–1421, http://epileptologie-bonn.de/cms/upload/homepage/weber/Fliessbach_Neuroimage_2006.pdf

*Foley, G., 2004, An Invitation to Systems Thinking: An Opportunity to Act for Systemic Change, LCWR Global Concerns Committee, Maryland, https://lcwr.org/sites/default/files/page/files/Systems_Thinking_Handbook.pdf

*Forrester, J. W., 1961, Industrial Dynamics, Cambridge, Mass., The MIT Press, http://books.google.com/books/about/Industrial_dynamics.html

*Forrester, J. W., 1968, Principles of Systems, Waltham, MA, Pegasus Communications, ISBN. 0915299879, 9780915299874, http://books.google.com/books/about/Principles_of_systems.html

*Forrester, J. W., 1971, World Dynamics, Wright-Allen Press, http://books.google.ca/books/about/World_dynamics.html?id=vrdEAAAAIAAJ

*Forrester, J. W., 1979, Urban Dynamics, MIT Press, ISBN. 0262060264, 9780262060264, http://books.google.com/books/about/Urban_dynamics.html

*Forrester, J. W., 1987, Lessons from System Dynamics Modeling, System Dynamics Review, Vol. 3 No. 2, pp. 136-149, http://onlinelibrary.wiley.com/doi/10.1002/sdr.4260030205/abstract

*Forrester, J. W., 1989, The Beginning of System Dynamics, International Meeting of the System Dynamics Society, Stuttgart, Germany, http://leml.asu.edu/jingle/Web_Pages/EcoMod_Website/Readings/SD+STELLA/Forrester-Begin'g-SD_1989.pdf

*Forrester, J. W., 1991, System Dynamics and the Lessons of 35 Years, A chapter for The Systemic Basis of Policy Making in the 1990s, ftp://nyesgreenvalleyfarm.com/documents/sdintro/D-4224-4.pdf

*Forrester, J. W., 1992, System Dynamics and Learner-Centered-Learning in Kindergarten through 12th Grade Education (D-4337), Cambridge, Mass., MIT,

http://www.mitocw.espol.edu.ec/courses/sloan-school-of-management/15-988-system-dynamics-self-study-fall-1998-spring-1999/readings/learning.pdf

*Forrester, J. W., 1994, System Dynamics, Systems Thinking and Soft OR. System Dynamics Review, Vol. 10, No. 2, http://sdg.scripts.mit.edu/docs/D-4405-2.SD.SysTh.SoftOR.pdf

*Forrester, J. W., 1996, Leadership in a Changing Society, Cultural Leadership Forum, Concord, Massachusetts, http://sdg.scripts.mit.edu/docs/D-4620-2.Leadership.Weeks.pdf

*Fuenmayor, R., 1997, Recovering Systems Thinking from Systems Thinking, Systemist, Vol. 19, No. 2, 1997, http://www.saber.ula.ve/bitstream/123456789/15924/1/fuenmayor-recovering.pdf

G

*Galileo, G., 1962, Dialogue Concerning the Two Chief World System, Translated by Drake, Berkeley, ISBN for 1967 print: 0520004507, 9780520004504, http://books.google.com/books/about/Dialogue_Concerning_the_Two_Chief_World.html

*Gharajedaghi, J., 2004, Systems Methodology, A Holistic Language of Interaction And Design, http://ackoffcenter.blogs.com/ackoff_center_weblog/2004/03/systems_methodo_1.html

*Giorbran, G., 2005, The Two Opposing Types of Order in Nature, http://macrocosmicsymmetry.com/pdf/two-orders.pdf

*Godino, J. D., …, Mathematical Concepts, Their Meanings and Understandings, Proceedings of XX Conference of the International Group for the Psychology of Mathematics Education. Vol. 2, pp. 417-425, http://www.ugr.es/~jgodino/articulos_ingles/meaning_understanding.pdf

*Goodman, M., 1991, Systems Thinking as a Language, The Systems Thinker Vol. 2, No. 3, Pegasus Communications Inc., http://thesystemsthinker.com/tstlang.html

*Gray, R. M., 2009, Entropy and Information Theory, Springer-Verlag, New York, http://ee.stanford.edu/~gray/it.pdf

H

*Hacker, P. M. S., 1982, Events and Objects in Space and Time, Mind, 1982, Vol. XCI, No. 361, pp.1-19,

http://mind.oxfordjournals.org/content/XCI/361/1.extract

*Hammond, D., 2005, Philosophical and Ethical Foundations of Systems Thinking, tripleC, Vol. 3, No. 2, 2005, pp. 20-27, https://dspace.jaist.ac.jp/dspace/bitstream/10119/3842/1/20204.pdf

*Hedge, A., 2011, Systems Thinking, Cornell University, http://ergo.human.cornell.edu/studentdownloads/DEA3250pdfs/systems.pdf

*Hitchins, D. K., ..., Putting Systems to Work, ..., http://www.behsad.com/Portal/Portals/3/MyFiles/e-Putting%20SystemsToWork.pdf

I
*Ison, R. L., 2008, Systems Thinking and Practice for Action Research, The Sage Handbook of Action Research Participative Inquiry and Practice (2nd edition), Sage Publications, UK, pp. 139–158, http://oro.open.ac.uk/10576/1/Ison.pdf

K
*Kay, J. J., Foster J. A., 1999, About Teaching Systems Thinking, Proceedings of the HKK Conference, pp. 165-172, University of Waterloo, http://www.trinitiesofwisdom.org/wp-content/uploads/2011/07/How-to-teach1.pdf

*Kevin, M. G., Adams, Mun, H. N., 2005, Towards a System of Systems Methodology Once Again, 26th ASEM National Conference Proceedings, 2005.

*Kirkwood, C. W., 1998, System Dynamics Methods: A Quick Introduction, College of Business, Arizona State University, http://nutritionmodels.tamu.edu/copyrighted_papers/Kirkwood1998.pdf

*Kline, J., 2006, Systematic Thinking, Primedia Business Magazines & Media Inc., http://www.jklineco.com/pdfs/operations_fulfillment_june_2001.pdf

L
*Lane A., 2008, Systems Thinking, Creative Commons, The Open University, UK, http://www.dialoguematters.co.uk/downloads/EA%20conf%20Systems%20thinking%20and%20ecosystems%20approach.pdf

*Larson, R., Hansen, D., 2005, The Development of Strategic Thinking: Learning to Impact Human Systems in a Youth Activism Program, Meetings of the Society for Research on Adolescence, Human Development, No. 48, pp. 327–349, 2005, S. Karger AG, Basel, http://youthdev.illinois.edu/Documents/LarsonHansen2005.pdf

*Larsson, M., 2009, Learning Systems Thinking: The Role of Semiotic and Cognitive Resources, Lund university, ISBN. 9197738069, 9789197738064, http://books.google.com/books/about/Learning_Systems_Thinking.html

*Laszlo, A., Krippner, S., 1998, Systems Theories: Their Origins, Foundations, and Development, Systems Theories and A Priori Aspects of Perception, Amsterdam, Elsevier, Science, 1998, Ch. 3, pp. 47-74, http://archive.syntonyquest.org/elcTree/resourcesPDFs/SystemsTheory.pdf

*Laszlo, E., 1972, The Systems View of the World: The natural philosophy of the new developments in the sciences, New York, Braziller, Update of this book is Laszlo, E. ,1996, The systems view of the world: A holistic vision for our time. Cresskill, NJ, Hampton Press, http://books.google.com/books/about/The_systems_view_of_the_world.html

*Laszlo, M., Lencse, G., 2008, Developing a Meta-methodology, http://www.hiradastechnika.hu/data/upload/file/2008/2008_7/HT_0807_2MukaLencse.pdf

*Lazanski, T. J., ..., Systems Thinking: Ancient Maya's Evolution of Consciousness and Contemporary Systems Thinking, University of Primorska, Portoroz, Slovenia, http://www.turistica.si/downloads/obvestila/NagradaJereLazanski/JereLazanski-AngleskaPrezentacija.pdf

*Longoria, R.G., 2008, Basic Feedback Control Concepts, Department of Mechanical Engineering, The University of Texas at Austin, http://www.docstoc.com/docs/48127617/Basic-Feedback-Control-Concepts

*Lopes, E., Bryant, A., ..., SSM: A Pattern and Object Modeling Overview, Overview of "Soft Systems Methodology (Checkland, P., 1993) and Object-Oriented Modeling and Patterns (Alexander et al. ,1997), http://www.leedsmet.ac.uk/ies/reddot/Eurico%20Lopes.pdf

M

*Maani, K. E., Maharaj, V., ..., Systems Thinking and Complex Problem Solving: A Theory Building Empirical Study, University of Auckland, New Zealand, http://www.systemdynamics.org/conferences/2001/papers/Maani_1.pdf

*Maier, M. W., 1998, Architecting Principles for Systems of Systems, Systems Engineering, Vol. 1 No. 4, pp. 267-284, 1998, http://onlinelibrary.wiley.com/

*Martínez-Vela, C. A., 2001, World Systems Theory, ESD.83 – Fall 2001, http://web.mit.edu/esd.83/www/notebook/WorldSystem.pdf

*Mathiassen, L., Munk Madsen, A., Nielsen, P. A., Stage, J., 1991, Soft Systems in Software Design, Systems Thinking in Europe, Plenum Press, http://www.larsmathiassen.org/12.pdf

*McLucas, A., ..., Systems Thinking and Decision Making: Risk Management and Situation Awareness, Argo Press, http://www.resources.nsw.gov.au/__data/assets/pdf_file/0007/261592/2---Alan-McLucas.pdf

*Meadows, D. H., Meadows, D. L., Randers, J., Behrens, W. W. I., 1972, The Limits to Growth, New York, University Books, http://www.amazon.com/exec/obidos/ASIN/0876631650/wikipedia08-20

*Mejia, D. A., ..., The System Idea and the Act of Knowing, The University of Hull, Bogota, Colombia, http://wwwprof.uniandes.edu.co/~jmejia/PDF/system_idea_and_knowing.pdf

*Monson, R. J., 2007, Systems Thinking: The Art of Seeing. NDIA Systems Engineering Conference, http://www.dtic.mil/ndia/2007systems/Thursday/AM/Track8/5617_1008_1211.pdf

N
*Neill, C. J., ..., Systems Thinking: Whole Solutions for Whole Problems, Penn State, Great Valley, http://www.personal.psu.edu/cjn6/Personal/Systems.pdf

*North, K., 2005, An Introduction to Systems Thinking, In Practice: A publication of the Savory Center. September/October 2005. No. 103, http://courses.umass.edu/plnt597s/KarlsArticle.pdf

O

*Okuribido, M., 2009, Efficiency is doing things right, Effectiveness is doing the right things, UNESCO 4th World Summit on Arts and Culture, Johannesburg, http://www.findthatpdf.com/search-14710866-hPDF/download-documents-okuribidopresentation-pdf.htm

*OMG-UML, 2000, UML Statechart Diagram, OMG-UML V1.3, 2000, http://santos.cis.ksu.edu/771-Distribution/Reading/uml-section3.73-94.pdf

*Ossimitz, G., ..., Teaching System Dynamics and Systems Thinking in Austria and Germany, University of Klagenfurt, Austria, http://wwwu.uni-klu.ac.at/gossimit/pap/ossimitz_teaching.pdf

P
*Paajanen, P., Kantola, J., Karwoski, W., Vanharanta, H., ..., Applying Systems Thinking in the Evaluation of Organizational Learning and Knowledge Creation, Systemics, Cybernetics and Informatics, Vol. 3, No. 3, pp. 79-84, http://www.iiisci.org/journal/CV$/sci/pdfs/P983028.pdf

*Partee, B. H., 1978, Fundamentals of Mathematics for Linguistics, Springer, ISBN. 9027708096, 9789027708090, http://books.google.com/books/about/Fundamentals_of_Mathematics_for_Linguist.html

*Partee, B. H., Meulen, A., Wall, R., 2006, Basic Concepts of Set Theory, Functions and Relations, Partee lecture notes, http://people.umass.edu/partee/NZ_2006/Set%20Theory%20Basics.pdf

*Patching, D., 1990, Practical soft systems analysis, Pitman, London, ISBN. 0273032372, 9780273032373, http://books.google.com/books/about/Practical_Soft_Systems_Analysis.html

*Peirce, C. S., 1878, How to Make Our Ideas Clear, Popular Science Monthly 12, Jan. 1878, pp. 286-302, Writings of Charles Peirce, Vol. 3, Indiana University Press, http://www.peirce.org/writings/p119.html

*Poppendieck, M., 2002, Principles of Lean Thinking, http://meidling.jvpwien.at/uploads/media/LeanThinking.pdf

R
*Regional Reliability Organization, 2005, Development of Interconnection-Specific Steady State System Models, Standard MOD-014-0, http://www.nerc.com/files/MOD-014-0.pdf

*Richmond, B., 1991, Systems Thinking: Four Key Questions, High Performance Systems Inc., http://www.iseesystems.com/resources/Articles/ST%204%20Key%20Questions.pdf

*Richmond, B., 1993, Systems thinking: Critical Thinking Skills for the 1990s and Beyond. System Dynamics Review, Vol. 9, No. 2, pp. 113-133, 1993, http://www.clexchange.org/ftp/documents/why k12sd/y_1993-05stcriticalthinking.pdf

*Richmond, B., 1994, System Dynamics/Systems Thinking: Let's Just Get On With It, International Systems Dynamics Conference, Scotland, Lyme, NH, High Performance Systems Inc., http://onlinelibrary.wiley.com/doi/10.1002/sdr.4 260100204/pdf

*Richmond, B., 2001, An Introduction to Systems Thinking, High Performance Systems, Inc., http://www.iseesystems.com/resources/Articles/STELLA_IST.pdf

*Robert J. Allio, 2003, Russell Ackoff Iconoclastic Management Authority, Advocates a "systemic" approach to innovation, Strategy & Leadership, Vol. 31, No. 3, pp. 19-26, 2003, MCB UP Limited, http://www.acasa.upenn.edu/p19.pdf

*Rodgers, C., 2008, An Introduction to Systems Thinking, Vanguard Scotland Consultant, http://www.systemsthinkingmethod.com/resour ces/general/Intro_systems.pdf

*Rufat-Latre, J., 1994, Strategy and Systems Thinking Through Dynamic Storytelling, 1994 International Systems Dynamics Conference, http://www.systemdynamics.org/conferences/1 994/proceed/papers_vol_2/rufat045.pdf

*Rutherford, J., Ahlgren, A. (AAAS: American Association for the Advancement of Science, Project 2061),1994, Science for All Americans, Oxford University Press, ISBN. 0-19-506770-3, 0-19-506771-1, http://www.project2061.org/publications/sfaa/o nline/sfaatoc.htm

S
*Sanford, C., 2004, A Theory and Practice System of Systems Thinking: With an Executive's Story of the Power of Developmental and Evolutionary Systems Thinking, InterOctave Development Group Inc.

*Senge, P., 1990, The Fifth Discipline, Currency Doubleday, New York, ISBN for 2006 print: 0385517254, 9780385517256,

http://www.4grantwriters.com/Peter_Senge_Th e_Fifth_Discipline_1_1_.pdf

*Shah, N. B., Rhodes, D. H., Hastings, D. E., 2007, Systems of Systems and Emergent System Context, MIT, Proceedings CSER, http://web.mit.edu/adamross/www/SHAH_CSE R07.pdf

*Shalaer, S., Mellor, S., 1992, Object lifecycles: modeling the world in states, Yourdon Press, ISBN. 0136299407, 9780136299400

*Sterling, S., 2003, Whole Systems Thinking as a basis for Paradigm Change in Education, University of Bath, http://www.bath.ac.uk/cree/sterling/sterlingthe sis.pdf

*Sveiby, K. E., 1994, What is Information?, http://www.sveiby.com/articles/Information.ht ml

*Sweeney, L. B., 2001, Systems Thinking: A mean to understanding our complex world, Pegasus Communications, http://www.pegasuscom.com/pdfs/systems-thinking-introduction.pdf

*System Dynamics Group, 1999, Guided Study Program in System Dynamics, MIT Sloan School of Management, http://ocw.mit.edu/courses/sloan-school-of-management/15-988-system-dynamics-self-study-fall-1998-spring-1999/assignments/assn10.pdf

T
*Tarski, A. ,1956, The Concept of Truth in Formalized Languages, in Logic, Semantics, Meta-mathematics, Clarendon, Oxford.

*Tidd, J., 2006, A Review of Innovation Models, Imperial College, London, http://www.emotools.com/static/upload/files/in novation_models.pdf

*Tollin, N., ..., Systems Thinking and System Dynamics in Evaluation of Sustainability, UNESCO Chair of Sustainability at Technical University of Catalunya, Spain, http://www.wu.ac.at/inst/fsnu/vienna/papers/t ollin_et_al.pdf

W
*Whelan, J. G., 1994, Modeling Exercises, MIT, http://ocw.mit.edu/courses/sloan-school-of-management/15-988-system-dynamics-self-study-fall-1998-spring-1999/readings/modeling2.pdf

*Wiener, N., 1948, Cybernetics: Control and Communication in the Animal and Machine, Cambridge, Mass, ISBN for 1961 print: 026273009X, 9780262730099, http://books.google.com/books/about/Cybernetics.html

*Williams, B., 2005, Soft Systems Methodology, The Kellogg Foundation, http://www.kapiti.co.nz/bobwill/ssm.pdf

*Williams, B., Harris, B., 2005, Systems Dynamics Methodology, WK Kellogg Foundation, http://users.actrix.com/bobwill/AESSD.pdf

*Wolstenholme, E., 2004, Using Generic System Archetypes to Support Thinking and Modelling, System Dynamics Review Vol. 20 No. 4, pp. 341-356, 2004, http://onlinelibrary.wiley.com/doi/10.1002/sdr.302/abstract

*Wolstenholme, E., 2005, The Potential of System Dynamics, Leading Edge, Oct. 2005, No. 10, NHS Confederation, UK, http://www.iseesystems.com/resources/Articles/leading_edge_10.pdf

Dictionaries & Encyclopedias
*American Heritage Dictionary of the English Language, http://ahdictionary.com/

*Britannica Online Encyclopedia, http://www.britannica.com/

*BusinessDictionary.com, http://www.businessdictionary.com/

*Dictionary.com (A part of the IAC Corporation), http://dictionary.reference.com/

*Merriam Webster Dictionary (An Encyclopedia Britannica Company), http://www.merriam-webster.com/

*Oxford Dictionaries, http://oxforddictionaries.com/

*Stanford Encyclopedia of Philosophy, http://plato.stanford.edu/

*The Free Online Dictionary (By Farlex), http://www.thefreedictionary.com/

*Thesaurus (Dictionary.com), http://thesaurus.com/browse/

*Wikipedia (The Free Encyclopedia), http://en.wikipedia.org/wiki/Main_Page/

Websites
*12manage, http://www.12manage.com/

*Academic Leadership Journal, http://www.academicleadership.org/

*Ackoff Collaboratory, http://ackoffcenter.blogs.com/

*American Association for the Advancement of Science, http://www.aaas.org/

*Answers.com WikiAnswers, http://www.answers.com/

*Applied Systems Thinking, Michael Goodman and David Peter Stroh, http://www.appliedsystemsthinking.com/

*BioMed Central, http://www.biomedcentral.com/

*Creating Technology: Engineering and Biomedicine, http://www.creatingtechnology.org/

*DC Physics, http://www.dctech.com/physics/

*Dictionary and Encyclopedia Directory, http://www.wordiq.com/

*EZine Articles, http://ezinearticles.com/

*Free Management Library, http://managementhelp.org/

*Google, http://google.com/ http://translate.google.com/

*Hedonistic Imperative, http://www.hedweb.com/bgcharlton/

*Mental Model Musings, http://www.systems-thinking.org/

*Newton: Ask A Scientist!, http://www.newton.dep.anl.gov/

*Operating and Support Cost Analysis Model, http://www.oscamtools.com/

*Patterns in Nature, http://www.patternsinnature.org/

*Pegasus Communications: What is systems thinking? Systems Thinking in Action, Structure Causes Behavior, http://www.pegasuscom.com/catalog.html

*People and Place,
http://www.peopleandplace.net/

*Philosophical Connections,
http://www.philosophos.com/philosophical_con
nections/ThinkReliability

*Physical Geography,
http://www.physicalgeography.net/

*Principia Cybernetica Web, Bollen, J., Riegler,
A.,
http://pespmc1.vub.ac.be/

*Purdue University,
http://www.purdue.edu/

*Searchcio - Techtarget,
http://searchcio.techtarget.com/

*SERC: The Science Education Resource Center
at Carleton College,
http://serc.carleton.edu/

*Systems Thinking in Schools,
http://www.watersfoundation.org/

*The Open University,
http://www.open.ac.uk/

*The Purplemath Forums,
http://www.purplemath.com/modules/

*The Worksheets.Com,
http://www.theworksheets.com/

*Think Reliability,
http://www.thinkreliability.com/

*Thomistic Philosophy Page,
http://www.aquinasonline.com/Topics/

*University of Twente,
http://www.utwente.nl/cw/theorieenoverzicht/

*Various Things About Albert Einstein,
http://www.einstein-
website.de/z_information/variousthings.html.

*Vision Learning,
http://www.visionlearning.com/

System Societies
*American Society for Cybernetics,
http://www.asc-cybernetics.org/

*BCSSS · Bertalanffy Center for the Study of
Systems Science,
http://www.bertalanffy.org/

*Complex Systems Society,
http://cssociety.org/

*Global Association for Systems Thinking,
http://www.globalast.org/

*IEEE Control Systems Society,
http://www.ieeecss.org/

*IEEE Systems, Man, and Cybernetics Society,
http://www.ieee-smc.org/

*ISSS: International Society for the Systems
Sciences,
http://isss.org/world/

*Society for Organizational Learning,
http://www.solonline.org/

*System Dynamics Society,
http://www.systemdynamics.org/

*Systems Society of India,
http://www.sysi.org/

*The Cybernetics Society - UK,
http://www.cybsoc.org/

*The International Institute for Advanced
Studies in Systems Research and Cybernetics,
http://www.iias.edu/

*UK Systems Society,
http://www.ukss.org.uk/

www.ingramcontent.com/pod-product-compliance
Lightning Source LLC
Chambersburg PA
CBHW061153220326
41599CB00025B/4473